The CSI Construction Specifications Practice Guide

Construction Specifications Institute

WILEY

John Wiley & Sons, Inc.

This book is printed on acid-free paper. ♾

Published by John Wiley & Sons, Inc., Hoboken, New Jersey

Published simultaneously in Canada

Library of Congress Cataloging-in-Publication Data:

The CSI specifications practice guide / Construction Specifications Institute.
p. cm.
Includes bibliographical references and index.

ISBN 978-0-470-63520-9 (pbk.); ISBN 978-0-470-90485-5 (ebk);
ISBN 978-0-470-90486-2 (ebk); ISBN 978-0-470-90487-9 (ebk);
ISBN 978-0-470-95127-9 (ebk); ISBN 978-0-470-95146-0 (ebk)

1. Buildings—Specifications—Handbooks, manuals, etc. I. Construction Specifications Institute. II. Title: Construction Specifications Institute specifications practice guide.
TH425.C75 2011
692'.3—dc22

2010019503

Printed in the United States of America

10 9 8 7 6 5 4 3 2 1

Contents

Chapter 14 Federal Agencies 189

Chapter 15 Warranties 201

Chapter 16 Construction Bonds 213

Chapter 17 Construction Insurance 221

Chapter 18 Summary 239

References 241

Index 243

Introduction

Successful and cost-effective construction relies on appropriate communication of a project design by the architect/engineer (A/E) to the contractor and other project participants. From project conception through design and construction to facility management, effective communication of the project requirements depends largely on having complete and coordinated construction documents.

The responsibilities of the participants comprising the design team are determined by the project delivery method used. In the design-bid-build, design-negotiate-build, construction management, and owner-build methods of project delivery, the A/E executes the design phases and delivers to the owner the documents according to the owner-A/E agreement. Design-build project delivery differs in that the A/E is employed by the design-builder, and the design-builder executes both the design and construction. The term *A/E* will be used to identify the design professional responsible for the project design and construction documents, regardless of the project delivery method used.

This practice guide is based primarily on the traditional design-bid-build delivery method and variations are indicated for other delivery methods.

The term *contractor* might also refer to a construction manager as contractor (CMc) or a design builder.

Many product decisions are made throughout the development of the contract documents. Product representatives can be a part of those decisions by providing technical assistance to the A/E, owner, or contractor in the early stages of a project and continuing the assistance through the development of the contract documents. Product representatives should be familiar with the overall concepts of drawings and specifications and understand the implications of contract modifications. Thorough knowledge of all aspects of the represented product is essential when assisting the A/E with the development of the documents used for construction. Product representatives should present themselves to the A/E as technical experts and resources on the proper use and incorporation of reference standards, specifications, testing and certification, and applicable codes. The ability to review documents and advise the A/E on a cost-effective method of installation, or a new product that would contribute to the project's success, is a valuable service and can build credibility.

Construction documents define the rights of, responsibilities of, and relationships among the parties. Of these documents, the *Agreement* and the *General Conditions of the Contract* have been developed by professional associations such as the American Institute of Architects (AIA), the Engineers Joint Contract Documents Committee (EJCDC), and the Design-Build Institute of America (DBIA) into standard documents. Several different standard documents have been published by these organizations for the various project delivery methods and basis of payment customarily used for construction. The advantage of a standard document is that it provides familiarity through repeated use and also that this repeated use has resulted in clear and well-coordinated documents. This standardization has been extended to the project specifications and drawings as well.

The Construction Specifications Institute's (CSI) and Construction Specifications Canada's (CSC) *MasterFormat*® presents a master list of numbers and titles for divisions and sections, providing a uniform location for administrative, procedural, and product information for any type of construction project.

Similarly, *SectionFormat*™ and *PageFormat*™ set up standard formats for the presentation of requirements within a specification section and on each specification page. *UniFormat*™ provides a standard for organizing information on construction systems, assemblies, and elements. The U.S. National CAD Standard for Architecture, Engineering, and Construction (NCS), including the Uniform Drawing System (UDS), provides a standard for defining and locating information on the drawings, including the identification of standard symbols, abbreviations, and notations. Through the use of such standardization, information can be placed properly in a predetermined location for each construction project, thereby making the task of using the documents much easier.

Specifications can be more easily coordinated with the drawings, specifications sections can be coordinated with each other, and A/E consultants can correctly integrate their work with less effort and error. It is not practical to include sufficient notes on construction drawings to describe in complete detail all of the product qualities and installation methods. Separate written descriptions, referred to as *specifications,* more effectively communicate this type of information.

The drawings and specifications are complementary. Proper methods of writing specifications or developing master guide specifications must be employed. The information must be clear to the user and presented in a concise manner, and it must also be technically correct and complete.

Specifying methods include descriptive, performance, reference standard, and proprietary. The A/E or specifier must understand the preparation of procurement documents and have an understanding of contract conditions including construction insurance, warranties, and bonds. The preparation of conventional specifications, as well as particular procedures such as specifying for the purchase of goods or for public agencies or the preparation of shortform specifications, requires an understanding of construction documents.

Construction documents are defined as the written and graphic documents prepared or assembled by the A/E for communicating the project design for construction and administering the construction contract. Construction documents are also used to obtain permits from the authority having jurisdiction (AHJ). Providing complete, coordinated, and easily readable construction documents that demonstrate code compliance can assist in problem-free AHJ reviews and inspections.

A standardized approach to the location of project subject matter for both written and graphic documents greatly simplifies the retrieval of information and reduces the possibility of conflicts, discrepancies, errors, and omissions. Written project requirements are organized in an orderly fashion, following the project manual concept. A properly assembled project manual with each document in its assigned location will facilitate checking and coordination. Likewise, a drawing set organized in a standard format with uniform location for information will be easier to coordinate with the project manual and itself.

Another important principle governing the production of construction documents is that each document, written or graphic, has a specific purpose and should be used for that purpose. Each requirement should be stated only once and in the correct location. Information in one document should not be repeated in other documents.

Whether a project is simple or complex, the basic concepts for preparing, organizing, and coordinating construction documents are applicable. Construction documents include two major groups of documents: procurement documents and contract documents.

Preface

Introduction to the Practice Guide Series

Beginning with the publication of the first *Manual of Practice* (MOP) in 1967 continuing through the publication of its successor document, the *Project Resource Manual* (PRM), it has been the intent of the Construction Specifications Institute (CSI) that these publications embody accepted standards for the preparation of construction specifications and project manuals, and a detailed source of information on quality documentation for the life cycle of a facility.

Through these publications, CSI has sought to aid owners, designers, specifiers, contract administrators, contractors, construction product representatives, and facility managers in the performance of their jobs.

In 2008, CSI began an effort to update the knowledge formerly contained in the MOP and PRM to present it anew and ensure its continued relevance. As with the earlier collections of this knowledge, the intent is to provide an authoritative resource on the organization, preparation, use and interpretation of construction documents, encompassing the entire life cycle of a facility from conception through facility management.

To accomplish this update CSI established the Practice Guides Task Team. One of the task team charges was to organize the presentation of this information into modules to support areas of practice where CSI currently offers certificates and certifications, such as Project Delivery addressed by the Construction Documents Technology (CDT) certificate, Specifications, addressed by the Certified Construction Specifier (CCS) certification, Contract Administration addressed by the Certified Construction Contract Administrator (CCCA) certification, and Product Representation addressed by the Certified Construction Product Representative (CCPR) certification, as well as other areas of practice for which education and certification may be developed.

To keep current with changes in the industry, the Task Team also reviewed other CSI documents and standards, and updated references to them that appear in the Practice Guides. A similar effort was made to incorporate changes in contract documents produced by The American Institute of Architects (AIA) and the Engineers Joint Construction Documents Committee (EJCDC), and to introduce the new standard contract documents developed by the ConsensusDOCS Coalition.

The Task Team also recognized the growing impact of "green" or sustainable practices on the subject matter contained in the Practice Guides. Each Practice Guide now addresses the topic of sustainable practice to some degree while a more detailed examination of the topic is planned for a future Sustainable Practice Guide.

Two other topics that have had an impact on the Practice Guides are: Building Information Modeling (BIM) and Integrated Project Delivery (IPD). The growing impact of BIM on the practice of specification writing and its potential impact on quality documentation made a discussion of this topic imperative. Likewise IPD has grown in importance over the past several years and has had an impact on the way practitioners relate to the process of creating and interpreting construction documents.

The Practice Guide Series is not intended to be composed of static documents but to be a living set of guides with the capacity to change and be updated as the construction industry changes around them. The input of users of this Series will be critical to the future updating of the Series and the authors and reviewers welcome feedback from users.

Description of *The CSI Construction Specifications Practice Guide*

The CSI Construction Specifications Practice Guide describes the process needed to effectively communicate a project design from the architect/engineer to the contractor, as well as other participants involved in the process of project conception, design, construction, and operation. The successful communication of a project design depends on the creation of a complete and coordinated set of construction documents both graphical and written. *The CSI Construction Specifications Practice Guide* serves as a guidebook to the preparation, organization, and coordination of construction documents and focuses on the written description of a project and how that description relates to the other construction documents. *The CSI Construction Specifications Practice Guide* also describes the best practices for product selection and defines the roles and responsibilities of the various parties to the process. Whether the reader is a seasoned practitioner or just getting started in creating or using written construction documents, the *CSI Construction Specifications Practice Guide* provides the information needed to enable effective communication among the various parties involved in a construction project.

Additional CSI publications that complement the Practice Guides are available for download to purchasers of the Practice Guides. The following documents can be obtained at www.wiley.com/go/csipracticeguides:

- MasterFormat® numbers and titles
- UniFormat™
- SectionFormat™/PageFormat™
- Sample CSI Forms
- GreenFormat™ questionnaire
- Practice Guide Glossary

To access the content on this site, you will need the unique Access Code printed on the card included with this book. If you have purchased an ebook version of this title, please contact Customer Care at 877-762-2974 or at http://support.wiley.com for a unique Access Code in order to take advantage of the web site.

Acknowledgments

No project of this magnitude can be accomplished single-handedly. In the best CSI tradition many people volunteered to write, review, and contribute to the production of the CSI Practice Guides. CSI wishes to recognize the following people who contributed in many ways to this project. If we omitted anyone's name it was inadvertent not intentional.

Ross G. Spiegel, FCSI, CCS, CCCA, FAIA, LEED AP BD+C
Chairman, Practice Guides Task Team

Author of *The CSI Construction Specifications Practice Guide*:
George Wade Bevier, CSI, CCS, SCIP, LEED AP

Practice Guide Task Team Members:
George Wade Bevier, CSI, CCS, SCIP, LEED AP
Ellen Kay Crews, FCSI, CCS, FMP
Kathryn A. D'Andrea, CSI, CCS
Dennis J. Hall, FCSI, CCS, CCCA, FAIA, SCIP
Gregory J. Marking, FCSI, CCS, CCCA, AIA, NCARB, LEED AP
Wiley N. McMillan, Jr., FCSI, CCPR
Chris Pechacek, CSI, LEED AP
Jim Rains, CSI, AIA
James M. Robertson, FCSI, CCS, FAIA, NCARB
Walter R. Scarborough, CSI, CCS, CCCA, AIA, SCIP
Paul W. Simonsen, RA, FCSI, CCS, CCCA, LEED AP, NCARB, SCIP
Paul M. Sternberg, CSI, CCS, CCCA, AIA, LEED AP, NCARB

CSI Certification Committee Members 2008–2010
Ann G. Baker, RA, CSI, CCS, CCCA, LEED AP, SCIP
Peter A. Baker, CSI, CCPR, RCI
Joseph Berchenko, CSI, CCS, AIA
Bill Coady, CSI, CCPR
Sheryl L. Dodd-Hansen, FCSI, CCS, CCCA
George A. Everding, CSI, CCS, CCCA, AIA, LEED AP
Raymond E. Gaines, FCSI, CCS, AIA
Richard G. Howard, CSI, CCS, AIA, NCARB, SCIP
Gregory J. Markling, FCSI, CCS, CCCA, AIA, LEED AP, NCARB
Wiley N. McMillan, Jr., FCSI, CCPR
Kay M. Melcher, CSI, CCS, LEED AP
Jonathan M. Miller, FCSI, CCS, CCCA, AIA, SCIP
Stephen E. Nash, CSI, CCS, CCCA, AIA, LEED AP
George Rosamond, FCSI, CCS, AIA, LEED AP
Marilyn G. Smith, CSI, CCS, AIA, SCIP
Robert B. Swan, CSI, CCS, CCCA, AIA
Robert W. Teller, FCSI, CCS, CCCA, LEED AP
J. Peter Tolson, CSI, CCS, CCCA
Lisa Turner, CSI, CCPR, LEED AP

Sandra M. Velleca, FCSI, CCS, CCCA
M. Keith West, CSI, CCPR
Laurence E. Wightman, III, CSI, CCPR, AHC

CSI Education Committee Members 2008–2010
Mary-Alice Avila, RA, CSI, CCS, LEED AP, MS
Loren R. Berry, FCSI, CDT, AIA, NCARB
Marc C. Chavez, CSI, CCS, CCCA, AIA, SCIP
Gregory J. Markling, FCSI, CCS, CCCA, AIA, LEED AP, NCARB
Robert T. Matschulat, CSI, CCS, AIA, AIA
Jack P. Morgan, RA, CSI, CCS, CCCA, AIA, NCARB
Stephen E. Nash, CSI, CCS, CCCA, AIA, AIA, LEED AP
David S. Proudfit, FCSI, CCS, AIA
Casey F. Robb, FCSI, CCPR, LEED AP
Paulette K. Salisbury, FCSI, CDT
Walter R. Scarborough, CSI, CCS, CCCA, AIA, SCIP
Gregory W. Sprinkel, FCSI, CDT
Grady W. Whitaker, Jr., FCSI
Jim F. Whitfield, FCSI, CCPR, CTC, LEED AP

CSI Technical Committee Members 2008–2010
Gary L. Beimers, FCSI, CDT
Lane J. Beougher, FCSI, CCS, CCCA, AIA, LEED AP George
Wade Bevier, CSI, CCS, SCIP, LEED AP
Tom R. Deines, FCSI, CCCA
Nina M. Giglio, CSI, CCS, SCIP
Mark J. Kalin, FCSI, CCS, FAIA
Michael MacVittie, CSI, AIA
Richard C. Master, CSI, AIA
Larry Nordin, CSI, CCS, AIA, ALA
James M. Robertson, FCSI, CCS, FAIA, NCARB
Paul M. Sternberg, CSI, CCS, CCCA, AIA, LEED AP, NCARB
David A. Stutzman, CSI, CCS, AIA, SCIP
Robert S. Weygant, CSI, CDT, SCIP

CSI Staff
Greg Ceton, CSI, CDT, Director of Technical Services

Chapter 1
Practices

1.1 Project Manual Concept

The documents commonly referred to as the *specifications* normally contain much more than that name implies. The procurement requirements, contracting forms, and conditions of the contract are usually included, but they are not specifications. In most cases, these are prepared by, or in coordination with, the owner and the owner's legal counsel and insurance adviser. This information, along with the specifications, is, in fact, a collection of certain written construction documents and project requirements whose contents and functions are best implied by the title "Project Manual." The project manual concept provides an organizational format and standard location for all of the various construction documents involved. For efficient coordination, document and section titles and their sequence of use should be the same for each project. With different methods of construction procurement, some documents become unnecessary. For example, in the owner-build delivery method and with negotiated contracts, bidding requirements are not applicable. Figure 1.1 shows an example of the recommended order of information and documents in a project manual. Separate volumes of the project manual are necessary when the number of pages makes them impractical to bind together in a single volume. Addenda and modifications are difficult to add to the bound project manual and are often compiled in a separate volume. Separate volumes can also be used for multiple contract construction projects.

1.2 Specifications Production

Specifications may be produced in a variety of ways, depending on individual office practice and, depending on the type of work generally performed, the number of specifications produced and the size of the architect/engineer's (A/E's) firm:

- In small firms, a principal assumes responsibility for specifications preparation.
- Medium-sized firms of 20 to 70 people might employ a full-time specifier.
- Large firms may maintain a complete specifications department of several people.
- A project-team approach may be used, with one person coordinating the writing efforts of several people working on the same project.
- An Internet-based consulting service may be retained on an as-needed basis.
- Any firm, regardless of its size and capacity, may elect to retain an independent specification consultant such as a member of the Specifications Consultants in Independent Practice (SCIP) organization on a project-by-project basis, or for all projects designed by the firm.

	DIVISION 00	**PROCUREMENT AND CONTRACTING REQUIREMENTS**
INTRODUCTORY INFORMATION	**00 01 01**	**Project Title Page**
	00 01 05	**Certifications Page**
	00 01 07	**Seals Page**
	00 01 10	**Table of Contents**
	00 01 15	**List of Drawing Sheets**
	00 01 20	**List of Schedules**
PROCUREMENT REQUIREMENTS	**00 10 00**	**SOLICITATION**
	00 11 13	Advertisement for Bids
	00 11 16	Invitation to Bid
	00 20 00	**INSTRUCTIONS FOR PROCUREMENT**
	00 22 13	Supplementary Instructions to Bidders
	00 30 00	**AVAILABLE INFORMATION**
	00 31 13	Preliminary Schedules
	00 31 16	Geotechnical Data
	00 31 19	Existing Conditions Information
	00 40 00	**PROCUREMENT FORMS AND SUPPLEMENTS**
	00 41 00	**Bid Forms**
	00 42 00	**Proposal Forms**
	00 43 00	**Procurement Form Supplements**
	00 45 00	**Representations and Certifications**
CONTRACTING REQUIREMENTS	**00 50 00**	**CONTRACTING FORMS AND SUPPLEMENTS**
	00 52 00	**Agreement Forms**
	00 60 00	**PROJECT FORMS**
	00 61 00	**Bond Forms**
	00 62 00	**Certificates and Forms**
	00 70 00	**CONDITIONS OF THE CONTRACT**
	00 72 00	**General Conditions**
	00 73 00	**Supplementary Conditions**
SPECIFICATIONS	**DIVISION 01**	**GENERAL REQUIREMENTS**
	DIVISIONS 02–19	**FACILITY CONSTRUCTION**
	DIVISIONS 20–29	**FACILITY SERVICES**
	DIVISIONS 30–39	**SITE AND INFRASTRUCTURE**
	DIVISIONS 40–49	**PROCESS EQUIPMENT**

Note: This example does not show a complete list of possible documents for inclusion in a project manual. See *MasterFormat*® for further information.

Figure 1.1
Project Manual Arrangement

- Building product manufacturers frequently provide technical services that include project-specific specifications for their product lines.
- Online building product services also offer free outline and short-form specifications generated from checklist selections by the designer or specifier.

In all instances, someone by choice or assignment will assume the role of specifier and will be responsible for assembling the project specifications. Like architectural designers, some of these specifiers may find that they have a predilection for or have an ability to be proficient at writing specifications.

1.3 Profile of a Specifier

Just as documentation associated with construction projects has become more specialized, so also have the qualities and characteristics of the specifier. Because specifiers produce written documents, they must be proficient in the use of written language. They must be capable of producing precisely worded sentences and well-constructed paragraphs that are clear, concise, and easily understood by a wide variety of people. Researching and reading skills are also of prime importance because of the requirement to work with voluminous quantities of information in search of precise information. In the course of preparing construction documents, specifiers consult and coordinate with other professionals in many disciplines. They must be skilled in verbal communication and credible in all of their dealings with others. Because they deal with construction methods and materials, the specifier must have and continue to develop a thorough working knowledge of construction products and materials, the construction industry, and field construction practices. Because construction contract documents are legal instruments, specifiers must also have a clear, fundamental understanding of the legal principles involved in each of the documents they prepare.

1.3.1 Educational Background

There is no single educational program designed to train a professional specifier. Any formal educational background must be supplemented with additional education gained through work experience before one can become a fully qualified specifier. Following are several formal educational backgrounds that would be suitable starting points for specifiers:

- College degree in architecture or engineering
- Vocational school education in architecture, engineering, or construction technology
- High school education supplemented by on-the-job training in architecture or engineering

1.3.2 Beneficial Traits

Following is a list of general character traits considered to be most desirable for a specifier. These abilities should be acquired if they are not inherent in the individual:

- Observes and is able to work with detailed information
- Organizes thoughts and concepts into orderly statements
- Can sort out the essential elements of a problem
- Is inquisitive, objective, and impartial
- Listens to and accurately understands people with various backgrounds, diverse objectives, and levels of authority
- Works with people and achieves equitable agreements without compromising important principles
- Accurately expresses ideas and concepts
- Can visualize three-dimensional objects

1.3.3 Proficiency

The effective specifier uses those character traits to develop the following areas of proficiency:

- Understanding of construction materials, systems, and methods
- Verbal and written communication skills
- Understanding of graphic information presented in drawings, tables, and charts
- Ability to carry out successful negotiations and coordination
- Management and organization techniques needed to meet scheduled commitments
- Thorough and methodical research methods
- Knowledge of computers as they are used to access information and perform word processing
- Understanding of basic construction law
- Understanding of building codes and ordinances
- Knowledge of basic insurance and bonds as they relate to the construction industry
- Ability to assist with quality control and quality assurance

1.4 Developing Specifications

The techniques used in the preparation of specifications usually involve development of sections from product reference material or editing sections from master guide specifications. Regardless of who writes the specification sections or how they are developed, products must be researched, selected, evaluated, coordinated with other products, specified in a consistent and clear manner, and coordinated with the drawings.

1.4.1 Gathering Information

Before beginning to write a section of specifications, the specifier must have the necessary information pertaining to that product, equipment, system, or assembly. Two kinds of information are needed. The first is information regarding the specific project requirements that may be obtained from the following:

- Owner's specific requirements such as mandatory products, standard forms, and standard details
- The A/E design team's drawings
- A specifications notebook or checklist with recorded product selections by the designer
- A preliminary project description or outline specification prepared earlier
- Applicable laws and building codes

The second kind of information is reference material pertaining to products and construction methods applicable to the particular specification section. Sources include:

- Manufacturer's information, including product data available on company web sites, hard-copy product data binders, manuals, catalogs, and other literature.
- Handbooks, pamphlets, and other educational and reference material published by trade associations whose members are manufacturers, fabricators, and installers.
- Reference standards available from trade associations, government agencies, and national standards associations.
- Information available from technical and professional societies such as the American Institute of Architects (AIA); the American Society of Civil Engineers (ASCE); the American Society of Mechanical Engineers (ASME); the American Society of Heating, Refrigeration, and Air-Conditioning Engineers (ASHRAE); Construction Specifications Institute (CSI); and the National Society of Professional Engineers (NSPE).
- Commercial master guide specifications such as AIA/ARCOM Architectural Computer Services, Inc. MasterSpec, InterSpec e-SPECS, Construction Sciences Research Foundation (CSRF) SPECTEXT, and Building Systems Design (BSD) SpecLink Suite.
- Product representatives who supply technical information and expertise.
- Information available from contractors, subcontractors, and special consultants.
- Personal experience of the members of the project team.
- Specifications for similar construction projects where similar products and methods were used. Caution should be used when referencing similar projects, as products and methods may have changed during construction and the final record drawings and project notes (i.e., change orders and other work/scope changes) should also be referenced before using this method. If the required specification information is not available in office files or catalogs, request assistance from the manufacturer's representative.

Many manufacturers offer suggested proprietary or guide specifications to relieve the A/E from some of the tasks of researching and writing a particular specification section. These specifications must be reviewed carefully to determine whether they have biases and whether they accurately state the necessary information. Manufacturer's suggested specifications that provide product-specific performance and installation data are often made available to A/Es on their web sites and through the ARCAT, McGraw-Hill and Reed MANU-SPEC program. Although many design team and contractor team members are involved in selecting and purchasing products and systems, the person actually developing the project specifications is the most likely to request comprehensive technical assistance. Product representatives should be able to perform the following tasks:

- Provide generic descriptions of products and systems.
- Understand the work of other trades that affects the proper installation and performance of the product.
- Identify the reference standards that relate to the product.
- Describe related work associated with the product.
- Assist in determining system, assembly, or product options or accessories.
- Identify modifications necessary to integrate a system or product in a project.
- Answer questions concerning delivery, storage, and handling of products.
- Explain installation procedures.
- Explain product and system installations.
- Identify the required certifications.

- Discuss requirements for field quality control.
- Answer questions regarding composition and manufacturing processes to produce products.
- Provide a guide specification in a format compatible with the A/E's computer system.
- Provide information regarding codes and regulations related to the use of a product or process.
- Assist in obtaining authorities having jurisdiction (AHJ) approval for installation of a product or process.

1.4.2 Product Selection Decisions

All product selection decisions should be recorded, and the A/E must direct or be made aware of product selection decisions. This requires systematic and progressive compiling of information, beginning with the early design phases. Selecting products for construction is like selecting goods for any other purpose. Selection is based on evaluation of the products against specific criteria derived from the owner's requirements. Selected products must be coordinated among team members. Once generic products are determined, each type must be investigated for possible sources. If there is more than one manufacturer of a product, investigate several manufacturers to determine acceptability. Cost savings to the owner may be achieved by allowing competition among acceptable products and documenting the criteria in the specifications. Some jurisdictions require competition unless special "sole source" permission is granted.

There are several factors to consider in evaluating a product:

- *Product.* Determine whether the product is appropriate for the project in all respects, including its functionality and aesthetics. Determine its physical limitations and compatibility with other products to be used. Confirm that the product is readily available and to what extent it is warranted. Determine the product's life expectancy and its maintenance requirements. Verify that the manufacturer's state that the product complies with codes and regulations.
- *Manufacturer.* Because much of the information and advice regarding the product and its application must come from the manufacturer, the manufacturer must be reputable. Determine what technical services the manufacturer provides and whether these services are available locally. Determine whether shop drawings are provided and whether orders are delivered on time and supervised during installation. Consider how the manufacturer will handle failures and problems involving its products.
- *Installation.* Determine whether the manufacturer's directions for installation cover the particular conditions of the project and whether there are alternate methods that must be designated in the specifications. Consider availability of skilled workers and complexity of installation. Determine what the preparatory and finishing requirements are for the product.
- *Cost.* If a material is inexpensive, it does not mean that the installation is as well, or vice versa. Also consider maintenance and operational costs. Initial cost should not be the only factor in selecting a product.

Most specifiers develop their own methods of assembling and recording specification information. Common methods are:

SPECIFICATION MEMORANDUM

Project: ROBERTSON SQUARE Project Number: 0312

Division: 06-WOOD, PLASTICS AND COMPOSITES Section: 06 42 00 - WOOD PANELING

PANELING IN THE EXECUTIVE OFFICES AND THE LIBRARY WILL BE BOOKMATCHED WALNUT VENEER.

JIM MITCHELL 3/15/10

Project Architect Date

Figure 1.2
Sample Specification Memorandum

- *Notes or E-mail Memoranda by the Project A/E to the Specifier.* Some project A/Es pass product information to the specifier by memoranda, an example as shown in Figure 1.2. Product information memoranda or e-mails should be filed under the appropriate section within the *MasterFormat*® titles of the specifications. The minutes of various meetings may also provide information affecting the specifications.
- *Product Notebook.* In the process of preparing specifications, the A/E should refer to the product notebook for information and catalog cuts collected during the design stage. This notebook will serve as a compilation of products selected for the project, so there will be less need to query the project A/E about decisions already made.
- *Specification Checklists.* Some firms use a specification checklist as shown in Figure 1.3. These forms list items by appropriate section and serve as reminders to the A/E. A comprehensive specification checklist must be compiled gradually, based on experience. A checklist of the materials to be used for a project may be circulated among the design team for use in preparing drawings and making notations. The A/E will use these lists to collect information and to start any necessary research and may use a checklist as a basis for the specification notebook. Checklists will often prompt the project A/E to record the necessary information. This process results in thorough specification notes and less time spent collecting information. If a comprehensive specification checklist is not used, the A/E may prepare separate worksheets for each specification section. A list of specification items shown on the drawings should be recorded on these sheets. Each drawing should be reviewed separately, and information available from previous notes and from preliminary project descriptions or outline specifications should be added on the worksheets. If errors, conflicts, or omissions are detected, comments should be made on the drawings, in the notebook, or on the worksheets for later correction and coordination. A meeting among project participants is usually the appropriate method of resolving such problems, and decisions made should be promptly noted on the worksheets or specification notebook.

Sample Coordination Checklist:
Section 08 31 13—Access Doors and Frames

- Location and size of each access door and floor door required.
- Locations of fire-rated access doors and required fire-resistance rating.
- Construction details such as those for masonry, gypsum board assemblies, plaster, tile, and acoustical surfaces in which access doors will be installed.
- Schedule of access doors.
- Hand of the access door and clearance around doors.
- Details of nonstandard units that require custom fabrication.
- Identification of custom units and standard units fabricated from other than the usual materials or finishes.
- Locations of floor fire doors painted yellow with the following warning painted in black letters on the surface: FIRE DOOR—DO NOT STORE MATERIALS ON SURFACE.
- Details of safety railings on floor doors.

Figure 1.3
Sample Coordination Checklist

- *Preliminary Project Description (PPD).* These descriptions based on systems and assemblies may have been created for a project during schematic design and can be used as a checklist in preparing the project specifications.
- *Outline Specifications.* As done with PPDs, the outline specifications based on products and activities may also have been created for a project during design development. These may be more up to date and may also be used as a checklist in preparing the project specifications.

1.5 Specifications Organization and Preparation

Several decisions are required at the outset of the development of a specification section. On small, limited-scope projects, remodeling work, and work under the control of an owner-builder, the decision may be made to use a shortform specification or to record only the basic product selections and requirements. Conversely, highly detailed language may be needed for certain sections where requirements are critical or have been the subject of extensive investigation. Specific project requirements of the owner should be reviewed at this time, along with any requirements imposed by outside sources, such as regulatory agencies.

1.5.1 Basic Procedural Decisions

Before the specifier can begin editing master guide specification sections or writing original sections for a project, the type and organization of the specification must be decided upon. *The Uniform Location of Subject Matter* (EJCDC 1910–16), published by the Engineers Joint Contract Documents Committee (EJCDC), is a useful reference for coordinating portions of construction documents. For each subject that might be included in the construction documents, the recommended primary location, as well as any secondary locations for information, is provided. Suggested cross-references to other locations are also included. The information or requirements stated in secondary locations supplement rather than duplicate those stated in the primary location.

1.5.2 Format

Information collected should be developed in the three part *SectionFormat™*. This three-part format provides consistency in the presentation of information and at the same time allows flexibility for adapting to a wide variety of construction projects. To permit easy reference, only one specific requirement or action should be addressed in each article, paragraph, or subparagraph. If changes are necessary, items can be modified or deleted with minimum disruption of other text or cross-referencing.

1.5.3 Method of Specifying

A decision must also be made as to the appropriate method of specifying: descriptive, performance, reference standard, or proprietary. At times, more than one method may be necessary in the same project manual. The specifier should be careful about combining methods in a specification of a single product, as this can be redundant and can result in a conflict incapable of resolution. The specifier should evaluate the complexity of the specification required for the project. A project manual may contain a combination of sections that vary in their complexity. On small projects, remodeling projects, or even more substantial work under the control of an owner, the decision may be made to use a shortform specification with sections that contain the requirements of an entire division, recording only basic decisions, selections, and requirements. Conversely, highly detailed sections may be needed where requirements are complex, unique, or extensive. Some master guide specifications include sections of varying content. The most appropriate text or version for each section should be selected based on project requirements. The project requirements might also dictate the method or methods of specifying that can or should be used. If a project is for a government agency, nonrestrictive methods will generally be required, and it is essential to be aware of this requirement before beginning to prepare specifications.

1.5.4 Specification Language

Competent specification writing requires the correct use of vocabulary and grammar along with correct construction of sentences and paragraphs. Always use the four principles of effective communication:

- Clear
- Concise
- Correct
- Complete

The imperative mood should generally be maintained throughout a specification. Consistent use of terminology and language contributes to better communication. Avoid duplicating or contradicting requirements contained elsewhere in the project manual. Because the contractor and owner are the only parties to the owner-contractor agreement, all instructions are addressed to the contractor. The words "the contractor shall" are generally omitted but may be added for clarity when both parties to the contract are mentioned in the same article or paragraph. Do not address individual subcontractors or trades. References to a specific responsibility should be made to the specification section—for example, "Wiring to unit is specified in Section 25 06 00" rather than "Electrical subcontractor is responsible for wiring unit." Recognized technical dictionaries, trade associations, and American Society for Testing and Materials (ASTM) standards on terminology should serve as a guide for correct use of technical terms. Consistency in the use of terms will help minimize confusion and ambiguity.

1.5.5 Preparing Specifications

Specification sections can be newly written, derived from a previous project specification, or edited from a master guide specification. With the use of commercial master guide specifications, editing rather than writing is usually the primary means for producing project manuals or specifications. Copies of previous project specifications should be used only for reference and should not be edited for a new project. Previous specifications may not contain necessary options, materials, or standards and may have changed since the specification was written, or the product may no longer be available. In addition, copied sections might not reflect changes made by addenda or contract modifications.

1.5.6 Division 01—General Requirements

Early distribution of the Division 01 draft to consultants and other contributors to the specifications will help eliminate overlapping requirements and omissions.

1.5.7 Section Title List

Once basic requirements for the project are known, a list of section titles is developed, along with the scope of work associated with each. This list should also indicate related work that is specified in other sections. The section title list serves as a device for coordination among the sections and helps prevent overlaps or omissions in the specifications. *MasterFormat®* can serve as a checklist for developing the section title list and ensuring that all required sections are included. Because the list of sections is usually distributed to the owner and to other design team members, it is preferable to list other documents and forms that will be included in the project manual. Such a list in its final form can become the table of contents for the project manual. If a master guide specification is used, the list of proposed section titles should be compared with the table of contents of the master guide. Sections that can be developed or edited from the master guide text, as well as those that must be custom-prepared, should be indicated. Attention should be given to products or work that might be specified in more than one section. When a master guide specification is used, care should be taken when changing section numbers, as cross-references may be included in other sections of the master. Other sections of the specifications should be referenced as needed for proper coordination of the requirements. References should be made to section numbers rather than to article or page numbers, which may change with subsequent revisions of the specifications.

1.5.8 Coordinating with Consultants

Portions of the project specifications may be written by consultants. This is often the case for the mechanical, electrical, structural, landscape architectural, and civil engineering sections because these disciplines are normally required to professionally seal their documents. Other sections may also be written by specialists such as roofing and waterproofing, architectural hardware, commercial kitchen, theater consultants, and lighting designers. Such situations require special coordination among entities involved to ensure that the specifications are complete, compatible, consistent, and without duplication or overlap. The prime A/E should assume responsibility for overall coordination of the specifications. Information regarding the type of project, contract, specifications formats, general conditions, and especially Division 01 requirements should be made available to consultants prior to their preparing any sections. Input from them for Division 01 sections should be requested at the same time. In this manner the consultants can avoid writing separate general requirements for their disciplines, reducing the likelihood of conflicting requirements.

1.6 Specifying Workmanship

Workmanship is often confused with the term *quality*. As quality is a mirror of the requirements, the requirements need to reflect what is intended. *Workmanship* generally refers to precision and craftsmanship. These requirements have measurable properties and can be specified with a means to verify compliance. Workmanship can be controlled in several articles of a specification section. These articles specify qualifications, tolerances, and various other aspects of quality assurance and quality control. An excellent example of controlling workmanship is the *Architectural Woodwork Standards (AWS)*, jointly published by the *Architectural Woodwork Institute (AWI), Architectural Woodwork Manufacturers Association of Canada (AWMAC),* and *Woodwork Institute (WI)*. This publication establishes measurable tolerances for joinery, surfaces, and clearances, along with respective grades of premium, custom, and economy, which must be selected and specified. Workmanship can be divided into two main categories:

- Workmanship relating to the manufacturing and fabrication of products
- Workmanship relating to the application, installation, or erection of products

Workmanship relating to the fabrication should not be confused with the materials used in the product—that is, the fabrication of a product made of copper or stainless steel may not be any different from the fabrication of the same product made of carbon steel. Although materials have their own characteristics, they can be affected by the workmanship of both humans and machines. It is a combination of materials and workmanship that results in the final characteristics of the product. Characteristics such as flatness, reflectivity, and finish may result from the thickness and the fabricated techniques such as rolling or folding into certain shapes. The specified characteristics, including tolerances, will determine the requirements and thus the required quality.

Workmanship relating to the application and installation of products is equally important to ensure proper performance and aesthetically pleasing installation. It is in this area that the qualifications of the installer, applicator, and erector will control the final execution. The A/E's knowledge and experience can be of special value. A mock-up or field sample can be specified to establish an acceptable level of workmanship.

When workmanship is specified, care should be taken to distinguish between product and installation items. Each workmanship requirement is placed under the proper heading in the specification section: product workmanship in PART 2—PRODUCTS and installation workmanship in PART 3—EXECUTION. Product data usually contain adequate information for PART 2; however, the installation or application also needs special attention and can be of equal or even greater importance.

Construction projects are made from many interrelated products, each requiring specialized skills to make it a functional part of the project. Workmanship requirements should be specified to an appropriate level in accordance with the needs of the project. Project requirements should be specified without demanding conformance to unattainable standards or to standards exceeding those necessary for the requirements of the overall project.

Workmanship statements can be ambiguous if not properly worded. Workmanship requirements should be measurable. Avoid broad generalities such as "best possible workmanship." The word *best* is open to many interpretations by the contractors, subcontractors, and suppliers reading the specifications.

Appropriate methods for achieving desired workmanship include:

- Referring to applicable standards and codes
- Establishing qualifications of manufacturer, fabricator, and installer
- Referring to workmanship requirements of trade associations
- Requiring samples to establish an acceptable level of workmanship and a basis for judging subsequent work
- Specifying tolerances and performance or physical requirements
- Establishing testing and inspection requirements

1.7 Considerations during Development of Specifications

In addition to the technical and aesthetic aspects of products and materials along with their installation, performance, and longevity, the A/E should also keep in mind factors that may affect the cost and schedule of a project.

1.7.1 Economic Implications of Each Choice

To a large degree, it is left to the A/E to set the requirements for and, therefore, the cost of many items in the project. The A/E should be aware of the economic impact of the requirements stated in the specifications. Some A/Es fail to distinguish between the characteristics of a given material when used for different purposes. For example, the grade of redwood specified for an entrance door pull and the grade of the redwood strip indicated at the edge of the parking lot paving should not be the same. Avoid elaborate and expensive requirements for items that are not critical to the success of the project. Consider only the selection of appropriate grades of materials for the specific part of the project under consideration.

1.7.2 Work Results Structure

When organizing and writing specification sections, the A/E must understand that neither the section titles nor their arrangement are organized by work results and do not control how the work of the project is to be assigned to various construction trades and subcontractors. For example, the cold-formed metal framing for an exterior wall may involve welders and ironworkers, while non-load-bearing wall framing for gypsum board partitions may be erected by carpenters or lathers. Although *MasterFormat*® division and section titles may correlate with the activities of traditional contractor groups, and though the specification sections for products, materials, and systems installed by a single trade are often placed consecutively in its Master List of Numbers and Titles, it is not the intent of *MasterFormat*® to assign work to specific construction trades. Conversely, the subcontractor and the installing trade should not influence the arrangement of the specification sections. Published standard conditions of the contract further emphasize this philosophy within their requirements.

1.7.3 Industry Conditions and Availability of Products

The A/E should consider the availability of products being specified. A product available today may not be available some months later when the project is under construction. Similarly, some products require long delivery times and may not be appropriate for projects with short construction periods or when early completion of the project is extremely important. If the A/E is aware of items that may be susceptible to major availability or price fluctuations, these factors should be considered when making product selections. Another consideration is the availability of skilled workers within specific geographic areas. It may not be appropriate to specify a specialized construction technique for a project located in an area where skilled workers are not available to do the work.

1.8 Specifying Quality Assurance and Quality Control

The A/E prepares construction documents that set forth the requirements for construction of the project. These documents are based on the agreed-upon requirements, schedule, and budget. These requirements establish the project quality. The A/E's ability to administer quality assurance and quality control during construction depends on how clearly these requirements are stated in the contract documents. The contract documents establish requirements for the work and procedures for administering the contract. Specifications generally stipulate qualitative requirements, and the drawings generally indicate quantitative requirements. Therefore, the major criteria for quality are described in the specifications. Every specification requirement establishes the quality of work, which is not always limited to articles that use the word *quality*. Requirements range from procedural issues to performance criteria and workmanship.

1.8.1 Quality Assurance in the Conditions of the Contract

The basis for quality assurance is included in the conditions of the contract. For example, preconstruction submittals, such as bonds and certificates of insurance, are assurances of performance, payment, and financial responsibility. Most standard general conditions include warranty provisions requiring the contractor's affirmation that materials and equipment will be new and of good quality, unless otherwise required or allowed; will be free from defects not inherent in the quality required or allowed; and will conform to the requirements of the contract documents. The warranty provisions summarize the contractor's responsibility for quality and for conforming to the requirements of the contract documents. The warranty should not be confused with, and is not limited to, the one-year period for correcting defective work.

1.8.2 Quality Control in the Conditions of the Contract

The quality control provisions in the conditions of the contract require verification of conformance with the contract requirements. Many standard general conditions require the contractor to supervise the work using expertise, skill, and attention.

The contractor is solely responsible for controlling the means, methods, techniques, sequences, and procedures for construction and is therefore responsible for controlling the quality of the work.

The A/E usually visits the site to become familiar with the progress and quality of the completed work. On the basis of field observations, the A/E reviews the contractor's applications for payment and certifies or recommends payment of amounts due. The A/E through this process may alert the owner if the quality of work is found to be in nonconformance with the contract documents or below the level of completion represented by the payment request; however, primary responsibility of conformance with the contract documents still lies with the contractor. Standard general conditions typically give the A/E authority to reject work that is defective or does not conform to the contract documents. This authority does not constitute control of the work, nor does this authority establish a responsibility for means, methods, techniques, sequences, or procedures for construction. The A/E may require additional independent inspection and testing.

The general conditions typically stipulate that the contractor shall bear the expense of any additional testing and inspection only if the work is found to be defective. Work not conforming to the contract documents may be considered defective or nonconforming.

The owner, however, has certain rights and remedies with respect to defective or nonconforming work and may elect to accept such work along with an adjustment in the contract sum.

1.8.3 Division 01—General Requirements

Division 01 contains sections that specify procedures affecting project quality. The following are several Division 01 sections that often include provisions for quality assurance or establish quality controls over the work:

- *Regulatory Requirements.* Identifies AHJs along with related codes and requirements.
- *Project Management and Coordination.* Describes various types of meetings to ensure and monitor quality through interaction of the participants. A typical quality assurance meeting, which may be called for in an individual specification section, is the preinstallation meeting. All meetings, however, have the potential for some positive influence on quality.
- *Submittal Procedures.* Includes such things as shop drawings, product data, and samples and provide proactive quality assurance through mutual understanding of the design intent and the contract requirements. This documentation and the samples provide assurance that the required quality will be obtained.
- *Quality Assurance.* Includes subjects such as manufacturer qualifications, supplier qualifications, installer qualifications, and testing agency qualifications as well as field samples, mock-up requirements, and contractor's quality control.
- *Quality Control.* Includes subjects such as testing laboratory services, source and field testing, and inspection services.
- *Substitution Procedures.* Establishes procedures for proposing and evaluating substitutions during the bidding/negotiating/purchasing and construction periods without compromising quality.
- *Commissioning.* Identifies process and requirements for facility, system, and component operation testing and checkout.

1.8.4 Specifications

Divisions 02 through 49 are composed of individual sections, each addressing a distinct subject area. The issues and requirements concerning quality assurance and quality control specific to that section are addressed. The contract documents are so interrelated that none of them can stand alone. An examination of *SectionFormat™* will help clarify the interrelationships between the requirements of articles within individual sections in Divisions 02 through 49 and Division 01 specification sections. The requirements in Division 01 are dependent on provisions in the conditions of the contract.

PART 1—GENERAL of a specification section includes articles on administrative matters. Three articles that deal specifically with quality are "Administrative Requirements," "Submittals," and "Quality Assurance."

- *Administrative Requirements* include preinstallation meetings that may be required for complex items requiring coordination and understanding among several participants.
- *Submittals* include preconstruction submittals, shop drawings, coordination drawings, product data, samples, quality assurance and quality control submittals, informational submittals, construction photographs, and closeout submittals. Quality assurance and quality control submittals are generally in response to requirements

specified in the articles "Quality Assurance," "Source Quality Control," and "Field [or] Site Quality Control."

- *Quality Assurance Requirements* include prerequisites, standards, limitations, and criteria that establish an overall quality for products and workmanship and may include the following subjects:
    - *Regulatory Agency Sustainability Approvals* include procedures and requirements for posting bonds, jobsite and recycling requirements, restrictions on airborne pollution, requirements to submit a procedures plan, and securing review and approval by AHJ.
    - *Qualifications* state the minimum expertise of the manufacturer, supplier, fabricator, erector, applicator, or installer. They also include qualifications of testing agencies and a design engineer employed by the contractor or manufacturer.
    - *Regulatory Requirements* describe obligations for compliance with codes for contractor-designed items. They include the requirements of AHJs.
    - *Certifications* may be required to verify that performance is equivalent to that of tested units or specified criteria. Certifications may also be required by AHJs.
    - *Field or Site Samples* are usually used to state requirements for sample field applications of finishes, such as paint or other finish materials and coatings. Accepted samples may establish a minimum standard of workmanship.
    - *Mock-ups* state the requirements for full-size erected assemblies to ensure understanding and coordination of required construction. Mock-ups may also be used for testing and observation and for establishing standards by which workmanship will be judged.

PART 2—PRODUCTS of a specification section includes articles related to the manufacture and fabrication of products. Following are several articles that specifically address quality control:

- *Assembly or Fabrication Tolerances* establish a statistical range of acceptability for properties of products before they are incorporated into the project.
- *Source Quality Control* involves quality control of products during their manufacture. These controls become assurances of a product's quality.
- *Tests* may include compression tests for verification of a concrete mix design, sieve analysis of natural materials such as aggregates or soil, and test curves for fabricated units such as electrical switchgear.
- *Inspections* may include a checklist of items to be verified by inspection of products at their source. The qualifications of the inspector are significant when inspections require judgments based on experience and knowledge of a particular field.
- *Nonconforming Work* may include requirements for identification and correction of nonconforming work and final actions required to prepare installed products to perform properly.
- *Manufacturer Services* may require coordination during manufacture at the source for special inspections, including access by AHJ and other third parties.

PART 3—EXECUTION of a specification section covers work performed at the project site. The following articles in PART 3 relate to quality control:

- *Tolerances* is used to establish an acceptable range of deviation. Verification of dimensional tolerances may be required if the deviation appears unacceptable.

Dimensional tolerances may involve such issues as surface flatness or alignment. Frequency of the deviation from tolerances is sometimes controversial. A tolerance that indicates the deviation shall not exceed 6 mm in 3000 mm (1/4 inch in 10 feet) may be questioned as to direction and whether the deviation is cumulative, such as 20 mm in 9100 mm (3/4 inch in 30 feet); fragmentary, such as 6 mm in 300 mm (1/4 inch in 1 foot); or multiple, such as 6 mm every 150 mm (1/4 inch every 6 inches).

- *Field or Site Quality Control* represents the last form of verification and may form the basis for decisions about defective work during or after installation.
- *Field or Site Tests* usually involve quality control of variable conditions. Test methods, intervals between tests, and test sites are important issues. Field testing may include soil compaction, load tests, compression tests, and various other forms of nondestructive testing. Field testing may not always be performed at the site and may be performed on samples taken from the site.
- *Field or Site Inspections* may involve visual observation for compliance with specified criteria. This may involve such items as joints and connections.
- *Nonconforming Work* may include requirements for identification and correction of nonconforming Work and final actions required to prepare installed products to perform properly.
- *Manufacturer's Services* may require a manufacturer's representative to visit the site to instruct an installer or owner's personnel in the proper installation of a material or the operation of specialized equipment. The manufacturer's field service may also be required to verify compliance with the manufacturer's instructions. A field report is an appropriate quality control submittal for these services.

1.9 Preparing the Specifications

Specification sections may either be written from scratch following the CSI principles described in this chapter or be created by editing available master guide specifications. Although text for specification sections may be derived from several sources, the use of master guide specifications is most common. Preparing specifications from information that is not in a usable specification format should follow the principles and techniques detailed in this chapter.

1.9.1 Writing/Editing Preparation

The specifier should review project requirements for the work covered in each section. Preliminary project descriptions and outline specifications prepared during the early design phases may be helpful. It is also important to thoroughly review the drawings, details of construction, and schedules. Local conditions and governing codes must also be considered, as both proprietary and master guide specifications may not address local project conditions or may contain references that do not apply. Prepare a preliminary table of contents for the project manual by identifying the specification sections from *MasterFormat*® that will be needed to cover the extent of the project. As a general rule, sections of Divisions 02 through 49 should be written before completing Division 01 sections.

Careful coordination will be required if portions of the project manual will be written by others, such as consultants. The most effective writing sequence for an individual specification section is as follows:

1. PART 2—PRODUCTS is most frequently the first step in the writing process. Sometimes it is necessary to skip over secondary materials and fabrication methods until PART 3—EXECUTION has been written.
2. PART 3—EXECUTION normally follows, unless the installation process dictates type of materials. After PART 3 is completed, PART 2 should be checked.
3. Preparation of PART 1—GENERAL can be done most effectively after the substance of the other two parts has been established.

1.9.2 Specification Illustrations

Although graphic illustrations and drawings are usually included in the contract drawings, this does not preclude the use of illustrations within the specifications. The old cliché states "a picture is worth a thousand words," and one illustration can help minimize extensive descriptions and prevent misunderstanding. The illustrations can also assist the owner, contractor, and supplier in understanding the intended product. Typical items that would benefit from illustrations are hardware, toilet accessories, furniture, lighting fixtures, and many accessory items. Illustrations are sometimes presented as photographs or in an isometric view, which simulates a three-dimensional view and provides a good understanding. The use of illustrations, graphics, and photographs may require permission or copyright release from the owner or originator of the graphics. Some graphics may include proprietary or patented designs, which could imply additional specification requirements. Graphics should be used in specifications only with due consideration and text that clarifies the graphics' status within the documents. Such text may be included in the supplementary conditions under definitions of contract documents. Many manufacturers provide data and other informational sheets that can be included to illustrate the products being used as the basis of design in the project. They can be filed in a separate folder or binder to be provided upon request to the project and construction team. The data sheets can also be referenced in the individual section and the data or information sheet(s) can be inserted in the project manual following the individual section as suggested in *SectionFormat*™.

1.10 Producing a Project Manual

Preparations of project manuals vary with individual firm practice. In small firms, the principal may write or assemble the various written documents for the project manual. In larger firms, this work may be the responsibility of a specifier or project A/E. Regardless of who prepares the project manual, preparation should begin early in the construction documents phase. The following tasks should be performed, generally in this order:®

1. Obtain and review the owner-A/E agreement to understand the A/E's role and responsibility for construction documents, bidding or negotiation, and contract administration phases. Although the content of specification sections is largely

independent of contract forms, the A/E assembling the project manual will need to have an understanding of these requirements. The contract between the owner and the A/E should be reviewed to determine the extent of the A/E's involvement during the construction process. Such considerations may include review of submittals, certification for payment, meetings, and field observation. The specifications and the conditions of the contract should not require A/E services that have not been included in the owner-A/E agreement.

2. Request information from the owner for the type of construction contract, insurance, and bonds. Insurance requirements should be obtained from the owner or the owner's insurance adviser. *AIA Document G612, Owners Instructions to the Architect* and *EJCDC C-052, Owner's Instructions Concerning Bonds and Insurance for Construction*, can be used to obtain this information. It is also important to determine whether special testing and inspection services will be required and who will contract and pay for these services. Specific requirements and information furnished by the owner should be reviewed and coordinated with other requirements.
3. Review the proposed agreement between the owner and the contractor to understand the roles of the owner and the contractor during construction.
4. Review the proposed general conditions of the contract to understand the roles and responsibilities of A/E, owner, and contractor during construction. In some instances, the general conditions will be furnished or designated by the owner. Because the general conditions govern every specification section, the specifier must know which documents will be used. Master guide specifications rely on the use of a given set of general conditions and may contain references to them. Caution should be exercised in editing a master guide specification to avoid conflict with the general conditions. Additional editing is usually required to accommodate different conditions of the contract.
5. Prepare the supplementary conditions to modify the general conditions in order to coordinate the contract with the specific requirements of the agreement.
6. Prepare Division 01—General Requirements coordinated with the project requirements of the agreement and the conditions of the contract.
7. Send the proposed conditions of the contract and Division 01 to the owner and consultants for review. The consultants need to know the information in these documents in order to properly coordinate their documents without duplication or omission.
8. Prepare the project specifications. Divisions 02 through 49 specifications should be based on the project requirements.
9. Make decisions about materials, equipment, and methods.
10. Develop specifications or edit master guide sections.
11. Prepare or assemble other written procurement documents. In some instances, the owner will furnish instructions for procurement, procurement forms, and bid security forms. If so, they should be obtained at the beginning of the project in order to ensure proper coordination.
12. Coordinate graphic and written construction documents. Requirements from other sources must be identified, including those imposed by regulatory agencies. It is advisable to prepare a checklist of code requirements and local conditions. *AIA Document D200, Project Checklist*, may be helpful in compiling this information.
13. After final review, reproduce the project manual for distribution. Figure 1.4 describes the activities involved in developing a project manual.

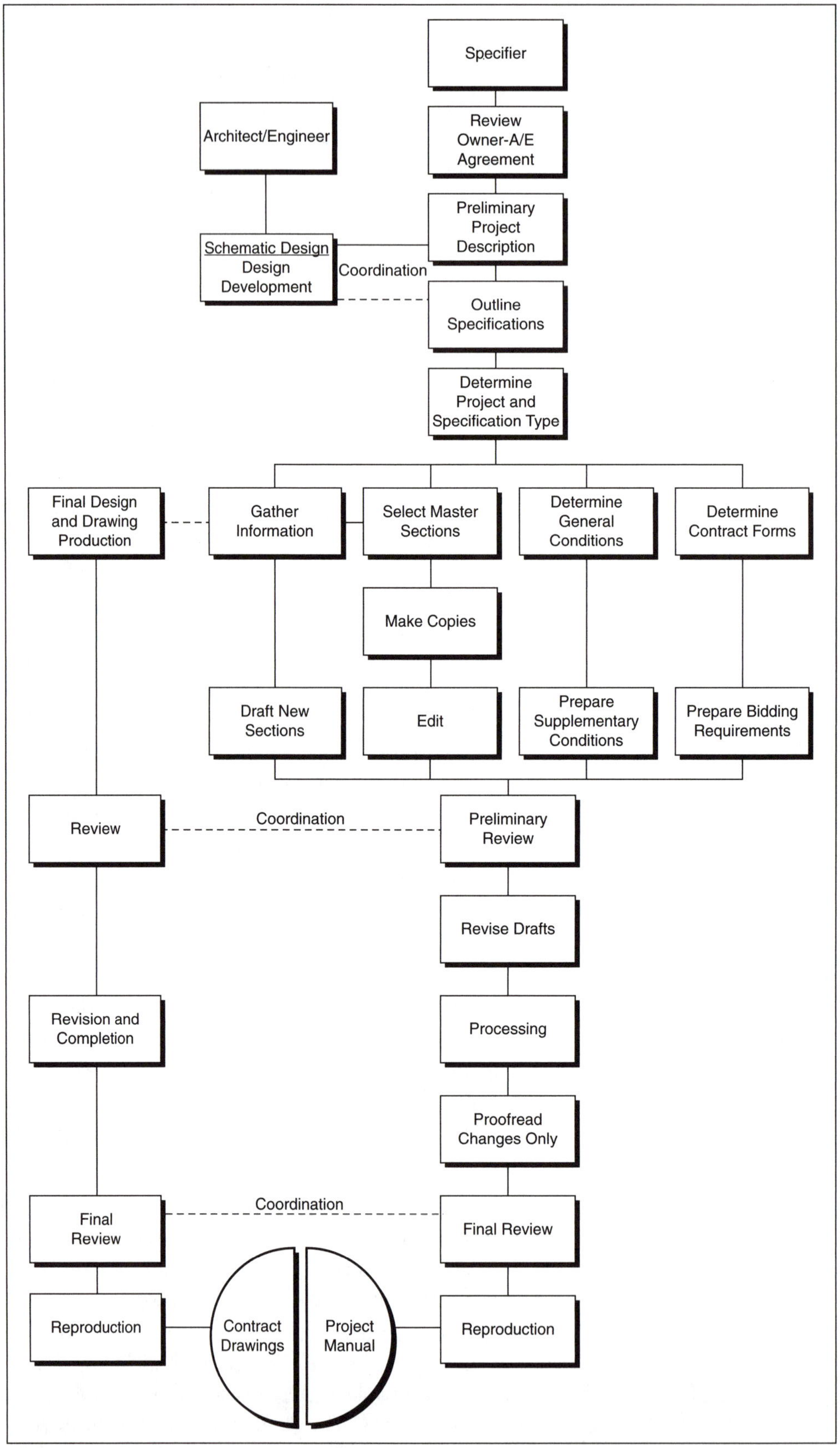

Figure 1.4
Project Manual Development

1.11 Coordination of the Specifications

Continuous communication and coordination of information among the project team members is essential to reduce errors, omissions, duplications, and inconsistencies within the project manual and with the construction drawings. Sharing of design and product decisions with project team members must begin at the earliest phases of the project, and the A/E should assume the responsibility for coordinating the flow and documentation of project information.

1.12 Preliminary Review and Coordination of the Project Manual

After the first draft of the specification has been prepared, a preliminary review and coordination of all the construction documents is performed. Copies of edited sections should be reviewed by members of the project team who are familiar with design and detailing to identify errors, oversights, and new or changed decisions. The content of edited master guide specifications should be reviewed by checking the range of alternatives and the selections that have been made. Drafts of sections from consultants should be collected and reviewed so the entire project manual can be coordinated. Sections should be checked and compared to eliminate omissions, overlaps, duplications, and inaccurate cross-referencing. Particular attention should be given to Division 01 sections. Where appropriate, procurement requirements should be prepared. These documents should be coordinated with applicable specifications sections. Alternates, allowances, and unit prices relate to the bid/proposal form and are cross-referenced to the instructions for procurement. The preliminary review of the project manual should also include coordination with the drawings. Consistent terminology, elimination of duplicate information, inclusion of all products, and other concerns should be checked. If the specifier discovers errors or omissions in the drawings, comments should be referenced to the drawings for corrections.

1.13 Final Preparation of the Project Manual

When no further substantial changes in content or extent are expected and project manual documents have been drafted, final text is processed from edited copy. Processing can be intended for producing final copies for the project manual or for final review and subsequent revision and reprocessing.

1.13.1 Proofreading

Proofreading should be performed by someone who is familiar with construction terminology and is competent to identify incorrectly selected editing options or statements out of context. The proofreader should review the selected options, revisions, and new paragraphs, sentences, statements, or sections that are added or deleted.

1.13.2 Table of Contents

The completed sections, schedules, reports, and other documents and forms are assembled for inclusion in the project manual. The project manual table of contents is prepared by carefully recording the section numbers and titles and, in some firms, the number of pages of each.

1.13.3 Reproduction and Distribution

The processed specification text should be transmitted to the owner for final review and comment. If changes are required, the text should be returned for revision. If changes are not required, the draft becomes the final copy for reproduction. Reproduce the accepted copy of the project manual and construction drawings in a number sufficient for distribution to each project team member for regulatory agency approvals, funding approvals, and procurement purposes and as a record document.

1.14 Variations Required by Project Delivery Methods

The preceding descriptions are based on the traditional design-bid-build project delivery method. Other project delivery methods require some variation in the construction documents. Some variations for project delivery may be simple, and others may require multiple project manuals and separate packages of drawings. Some variations will result from use of nonstandard agreements or conditions of the contract. Although multiple-prime contracts might be utilized without construction management, they are discussed under construction management project delivery in section 1.14.2. The procurement and contracting procedures may determine the type of documents required and the methods of specifying used. Information about contract procedures will typically be provided by the owner. *AIA Document G612, Owner's Instructions to the Architect,* provides three forms for obtaining the appropriate information: Part A for contracts, Part B for insurance and bonds, and Part C for bidding procedures.

1.14.1 Document Variations for Design-Negotiate-Build Project Delivery Method

The compilation of a project manual may be the continuation of negotiations that started during early design phases. The negotiations are frequently based on outline specifications and design development documents leading to a cost of the work with a guaranteed maximum price (GMP) agreement. The project manual and other construction documents are prepared to finalize the design, obtain required permits, and obtain actual prices for the cost of the work. These construction documents might then form contract documents. If the negotiating process does not begin until the design is complete, then the construction documents might be prepared for the entire project, including procurement requirements, to obtain proposals for negotiation. In this case, the variations may be limited to the procurement requirements.

Project Manual Variations There are some significant differences in project manual preparation between a design-negotiate-build and a traditional design-bid-build delivery method. The A/E will not necessarily be the sole decision maker on all the items that would normally be the A/E's responsibility. Decisions are often shared by other project team members.

Procurement Requirements *MasterFormat®* contains optional titles applicable to the negotiation process. These titles include Request for Proposal and Request for Qualifications, as well as Instructions to Proposers. Development of these documents can improve negotiations by establishing procedures for clarifications, an addendum, and a structured proposal form. These procurement requirements would establish similar requirements of a traditional bid process and provide affirmations such as affirming that the proposer has visited the site and accounted for local conditions. With a negotiated owner-contractor agreement, the contractor might prepare the contractor's own procurement requirements for subcontract packages.

Contracting Forms The owner-contractor agreement included in a project manual might be a standard form that will eventually include the negotiated amounts, resulting in the request for proposal (RFP), or might be determined by informal negotiations. The A/E might not always participate in this process. The A/E, however, should request a copy of the form of the owner-contractor agreement to ensure that there are no conflicts between the agreement and the conditions of the contract or Division 01. If the contract is negotiated between the owner and contractor, the A/E may not be involved in preparing contracting forms, bonds, and certificates. These forms might be prepared by the proposers. The A/E should review these forms to ensure that they do not contain responsibilities not included in the A/E agreement. The A/E should also inform the owner if the proposed agreement does not mention required responses to the A/E's responsibilities.

Conditions of the Contract Even though negotiations are the primary aspect of this project delivery, the conditions of the contract affect the various participants involved. The conditions of the contracts are the basis of the requirements established in Division 01. Conditions of the contract should be included in the project manual to establish this basis for the requirements. The general conditions might be standard documents published by the AIA or the EJCDC or might be general conditions offered by the proposer. Some owners, especially large developers, prefer to use their own general conditions, which are usually based on AIA or EJCDC general conditions. In some instances, the proposer offers the proposer's own version of the conditions of the contract, but these conditions should be reviewed by the owner's legal counsel. If the conditions of the contract are other than standard AIA or EJCDC general conditions, the A/E must carefully review the documents to understand their variations from the standard documents. Variations in general conditions might involve A/E and contractor roles different from those normally associated with standard agreements. These requirements or responsibilities must be reconciled. Requirements for insurance, bonds, and payments, including the process for payment applications and retainage, may be covered in the owner-contractor agreement and should not be repeated in the conditions of the contract. The owner-contractor agreement should be reviewed to ensure that conflicting requirements are not included in the conditions of the contract. If the owner-contractor agreement is not available to the A/E, the owner and contractor should prepare appropriate conditions of the contract for publication in the project manual.

Division 01—General Requirements Possible conflicts among the agreement, the conditions of the contract, and the Division 01 requirements can occur, depending on their source and particular requirements of the negotiations. Division 01 requirements should be prepared in conjunction with the conditions of the contract. Special coordination might be necessary for certain procedures that are to be negotiated, such as payment procedures, construction progress schedules, submittals, substitutions, warranties, and contract closeout.

Divisions 02 through 49—Specifications Writing specifications for this delivery method would not be significantly different. The decisions to utilize specific products might be the result of prior negotiations or value analysis; however, the specification would be similar. If pricing and negotiations have determined specific products, then the specification might be written as proprietary, eliminating a level of detail.

1.14.2 Document Variations for Construction Management Project Delivery

Construction management project delivery usually affects the construction documents when the construction manager is an adviser or agent. When the construction manager provides assistance during the design process, negotiates a contract, and then provides construction, the construction manager is a contractor. In either case, the nature of construction management is to provide management services and generally divide the project into various "packages" representing discrete scopes of work. These packages are generally competitively bid. Construction documents involving multiple-prime contracts may be released at different times and may require separate sets of documents, including a project manual and specific drawings. During the construction document stage of a construction management project delivery, the A/E may not prepare procurement documents. The Construction Manager as Adviser (CMa) may prepare these procurement documents for each bid package, including the bid solicitation, instructions to bidders, bid forms, and supplements, as well as a summary of work for each bid package or contract. If the A/E agreement was based on construction management services, the A/E will usually produce construction document packages identified by the CMa to implement the multi-prime contracts and fast-track techniques. Though the total project schedule may be reduced by utilizing these techniques, the overall document preparation time is usually longer. The construction documentation process is usually staggered to produce separate bid packages based on the required scheduling of the project.

Project Manual Variations Some differences in project manual preparation will occur using a CMa project delivery implementing multiple contracts. The A/E will not necessarily be the sole decision maker on all the items that would normally be the A/E's responsibility. Decisions are often shared with other project team members.

Division 00-Procurement and Contracting Requirements

Procurement Requirements The CMa may prepare the procurement requirements for the bid packages. Regardless of who prepares the procurement requirements, the A/E should work with the construction manager on the necessary drawings and specifications to be included in the packages and assist the construction manager in ensuring that the bid packages contain the required information. The process of obtaining separate bids is not significantly different from that of a traditional design-bid-build project delivery method.

Preparing procurement requirements for construction management project delivery can be an extensive process when multiple packages are involved. The A/E will be affected by how the construction manager chooses to issue bid packages. Some believe that full sets of procurement documents should be issued. The bid form, or Section 01 12 00—Multiple Contract Summary, should indicate the scope of work for each contract. The advantage to issuing full sets is that each bidder may examine all requirements that might affect its bid. The disadvantages, however, are that the design has to be complete and reproduction costs are high. A construction manager might issue only partial documents that describe the package being bid. The advantages are that bid packages may be issued while design is still under way, and reproduction costs are usually lower. The disadvantage is that bidders may not be able to examine all related documents.

Contracting Forms Agreement forms are usually prepared by the construction manager for each bid package. Some agreement forms used with construction management project delivery are as follows:

American Institute of Architects

- *AIA Document A132, Standard Form of Agreement Between Owner and Contractor, Construction Manager as Adviser Edition* (for use on projects where the basis of payment is either stipulated sum or cost of the work plus a fee, with or without a GMP)
- *AIA Document A133, Standard Form of Agreement Between Owner and Construction Manager as Constructor, where the basis of payment is the Cost of the Work Plus a Fee with Guaranteed Maximum Price*
- *AIA Document A134, Standard Form of Agreement Between Owner and Construction Manager as Constructor, where the basis of payment is the Cost of the Work Plus a Fee without a Guaranteed Maximum Price*
- *AIA Document B132, Standard Form of Agreement Between Owner and Architect, Construction Manager as Adviser Edition*
- *AIA Document B144/ARCH-CM, Standard Form of Amendment to the Agreement Between Owner and Architect, where the Architect provides Construction Management Services as an adviser to the Owner*

Associated General Contractors of America

- *AGC ConsensusDOCS 500, Standard Form of Agreement and General Conditions Between Owner and Construction Manager (GMP)*
- *AGC ConsensusDOCS 510, Standard Agreement and General Conditions Between Owner and Construction Manager (Cost of Work)*

Construction Management Association of America

- *CMAA Document A-1, Standard Form of Agreement Between Owner and Construction Manager, Construction Manager as Owner's Agent*
- *CMAA Document A-2, Standard Form of Agreement Between Owner and Contractor*
- *CMAA Document A-4, Standard Form of Agreement Between Owner and Designer*

- *CMAA Document CMAR-1, Standard Form of Agreement, Owner-Construction Manager ("CMAR" Series—CM At Risk)*
- *CMAA Document CMAR-2, Standard Form of Agreement, Construction Manager–Contractor*
- *CMAA Document CMAR-4, Standard Form of Agreement Between Owner and Designer*

Conditions of the Contract General conditions for construction management project delivery require unique provisions ensuring cooperation between and mutual responsibility among multiprime contracts. Special versions of standard documents are available and contain provisions for the CMa to provide services such as consolidating multiple applications for payment into a single application for the A/E to review. Some standard general conditions of the contract and bond forms generally used with construction management project delivery methods are as follows:

American Institute of Architects

- *AIA Document A232, General Conditions of the Contract for Construction—Construction Manager–Adviser Edition.* (When the construction manager is contractor (CMc), the standard version of AIA A201 is utilized.)

Construction Management Association of America

- *CMAA Document A-3, General Conditions Between Owner and Contractor*
- *CMAA Document CMAR-3, General Conditions, Construction Manager–Contractor*

Supplementary conditions may include the same modifications normally required in a traditional design-bid-build delivery method. These modifications include requirements for insurance and bonds and the process for payment applications and retainage. Because the conditions of the contract will be published with the first bid or contract package, early preparation is required. As with multiple prime contracts, identical conditions of the contract and the appropriate Division 01 sections should be issued with every package.

Division 01—General Requirements The A/E and construction manager must be alert to possible conflicts between the agreement, the conditions of the contract, and Division 01 requirements. Division 01 requirements for CMa (multiple-prime contracts) should be prepared with specific regard for the scope of each contract. Special coordination is necessary for certain procedures such as payments, progress schedules, product submittals, quality control, substitutions, warranties, and contract closeout. When separate bid packages are issued, not all Division 01 sections may be necessary for each bid package. Only the sections pertaining to the particular bid package should be issued with the respective package. For example, Section 01 45 36—Testing Laboratory Services would not be issued in bid packages that have no testing requirements. Certain Division 01 sections, such as allowances, alternates, testing laboratory services, temporary facilities and controls, field engineering, and operation and maintenance data should be issued only with the appropriate bid packages. The Division 01 sections common to all bid packages, such as submittal procedures, product substitution procedures, and cleaning, should be included with each bid package. The construction manager should be heavily involved in preparing Division 01 as part of the construction documents. The Division 01 sections, for construction management project delivery, inform the owner and the various prime contractors about administrative procedures required for the project. Division 01 may also be used to assign temporary construction requirements to various contractors. The following sections are generally included in CMa project delivery packages:

- *Section 01 12 00—Multiple Contract Summary.* Should identify the participants' responsibilities and long-lead items, bid packages, and use of the site.
- *Section 01 21 00—Allowances.* May be used to postpone design decisions. Allowances help the CMa achieve accurate cost control, even if some products are not known. These unknowns should be minimized because the CMa usually will have provided preliminary pricing. On multiple-prime projects, it is important to include the allowances in the appropriate contract.
- *Section 01 32 00—Construction Progress Documentation.* Includes requirements for critical path method network analysis with requirements for updating. Long lead items and milestone dates might be identified in the documents to ensure compliance with an overall project schedule.
- *Section 01 33 00—Submittal Procedures.* Should be included for consistency in the information required for submittals. The CMa field representative will usually maintain a complete set of submittals at the project site. The CMa will generally monitor submittals more closely to confirm scope of work and delivery dates. Shop drawings, product data, and samples are received and checked for completeness by the CMa and forwarded to the A/E or returned to the contractor for resubmittal.
- *Section 01 50 00—Temporary Facilities and Controls.* This section should coordinate site facilities and controls such as temporary utilities, parking areas, swing staging, field offices and sheds, sanitary facilities, hoists, telephones, fire protection, soil erosion, and other controls. Fences, barricades, walkways, project signs, and other minor temporary construction items should be also specified here along with responsibilities. Assignment of temporary construction to an appropriate trade contractor may be specified in this section. It is usually necessary to include specification requirements for temporary facilities and controls to allow the construction manager to divide these responsibilities among the contractors.
- *Section 01 70 00—Execution and Closeout Requirements.* Should be included for completing the CMa and contractor's duties, such as submission of warranties and maintenance data, as well as provisions for final cleaning and project record documents. The construction manager and the A/E may inspect the project for substantial completion. The results of this inspection are presented to the respective contractors for corrective action, when required, and to the owner for information.

Divisions 02 through 49—Specifications The A/E should prepare specifications to secure accurate contract bids. On multiple-prime contracts, the CMa may request changes to ensure clear definition of the responsibilities of various contractors. The level of detail of the specifications may be influenced by the CMa or by the owner's project requirements. For CMa (multiple-prime contracts), each bid package will usually consist of procurement requirements (including a bid form appropriate to the scope of the package), contracting requirements (including the conditions of the contract), and applicable specification sections. It is possible that a particular product specification section will be tailored differently for each bid package; therefore, a section with the same *MasterFormat*® number and title may vary slightly from one package to another. For example, Section 05 50 00—Metal Fabrications may contain only the miscellaneous metal related to each specific bid package. In this case, it is likely that several metal fabrication sections will be required for the entire project. Because each section will be different, each may be written as a Level 4 section and given a different section number and title (e.g., Section 05 50 00.10—Metal Fabrications for Building Envelope and Section 05 50 00.20—Metal Fabrications for Interior Construction). However, some construction managers may prefer to use the same product section for all bid packages and use Section 01 12 00—Multiple Contract Summary specification section to describe the scope of what is included in a particular bid package.

Combined Documents The owner or the CMa might require that all document packages be combined into one project manual for convenience. The A/E should be aware of this possibility from the beginning so that similar specification sections can be properly organized using the appropriate level of numbers and titles and the drawings can be numbered with a logical system.

1.14.3 Document Variations for Design-Build Delivery Method

In the traditional design-bid-build delivery method, the contract documents are a part of the contract between the owner and the contractor. In the design-build delivery method, however, the documentation prepared by the owner is the owner's project requirements or project description. The owner might issue an RFP, conditions of the contract between the owner-design-builder, technical specifications (Department of Defense) or performance specifications, and the design requirements (including any schematic drawings). The design-builder will respond to the RFP, negotiate an agreement, and eventually prepare the construction documents consisting of the specifications, drawings, and subsequent modifications. The construction documents may or may not be part of the contract documents between the owner and the design-builder, depending on the contract arrangements. However, they are usually submitted to the owner for approval. The construction documents for a design-build project are prepared by an A/E for the design-builder. Though performed in the interest of the design-builder, the basic services provided by the A/E may be similar to those of a traditional contract. In all cases, the construction documents should be prepared following principles in this chapter.

Division 00-Procurement and Contracting Requirements

Procurement Requirements The design-build contract between the owner and design-builder may be a negotiated contract resulting from an RFP from an owner. This process of selecting a design-builder may have occurred during the project delivery stage of the project. During this construction document stage, the design-builder will prepare any necessary procurement for subcontracts, including instructions to bidders. The subcontract procurement requirements and construction documents might not vary significantly from typical procurement documents prepared for the traditional design-bid-build delivery method.

Contract Requirements—Agreements and Conditions of the Contract Standard agreements and conditions of the contract between the owner and the design-builder for design-build projects are available from the AIA, EJCDC,

Design-Build Institute of America (DBIA), and Associated General Contractors (AGC). The design-builder may also have the design-builder's own general conditions tailored for a specific project type. Most standard agreements are written for two phases: the preliminary design phase and the construction phase. Most agreements incorporate the conditions of the contract into the form itself and therefore do not appear as separate documents.

AIA Document A141, Agreement Between Owner and Design-Builder, for use by an owner contracting with a single design-builder for both design and construction services, includes the agreement and three exhibits:

- The agreement involves review and evaluation of the program, budget, and site and submittal by the design-builder of design documents sufficient to establish the size, quality, and character of the project, along with the proposed contract sum and schedule. After written approval by the owner of the design documents, the agreement involves submittal of construction documents for review and approval of the owner and AHJ, as well as services necessary for the proper execution and completion of the work.
- The agreement also requires the parties to select the payment type from three choices: stipulated sum, cost of the work plus design-builder's fee, and cost of the work plus design-builder's fee with a guaranteed maximum price.
- Exhibit A includes the terms and conditions, while Exhibit B includes the determination of the cost of the work, and Exhibit C includes the insurance and bonds.

When the design-builder contracts with an A/E as a consultant, *AIA Document B143, Agreement between Design-Builder and Architect*, is available. This document consists of an agreement and two exhibits, coordinated with the agreement between the owner and the design-builder, one part for concept/preliminary design and the second part for construction documents. Exhibit A includes the initial information, and *Exhibit B, Standard Form of Architect's Services*, provides a menu of briefly described services that the parties can select and augment to suit the needs of the project

When the design-builder contracts with a contractor, *AIA Document A142, Agreement Between Design-Builder and Contractor*, is available for use. The form of agreement between contractor and subcontractors is *AIA Document A441, Standard Form of Agreement Between Contractor and Subcontractor for a Design-Build Project.*

EJCDC has different documents that cover preliminary services and a separate "owner's consultant" agreement in addition to the various design-build services. The documents are listed here in numerical order:

- *EJCDC D-001, Guide to Use of EJCDC Design-Build Documents*
- *EJCDC D-500, Standard Form of Agreement Between Owner and Owner's Consultant for Design Professional Services on Design-Build Projects*
- *EJCDC D-505, Standard Form of Subagreement Between Design-Builder and Engineer for Design Professional Services*
- *EJCDC D-510, Standard Form of Agreement Between Owner and Design-Builder for Preliminary Services*
- *EJCDC D-520, Standard Form of Agreement Between Owner and Design-Builder, Stipulated Price*
- *EJCDC D-521, Suggested Form of Subagreement Between Design-Builder and Subcontractor, Stipulated Price*

- *EJCDC D-525, Suggested Form of Agreement Between Owner and Design-Builder on the Basis of Cost-Plus*
- *EJCDC D-526, Suggested Form of Subagreement Between Design-Builder and Subcontractor, Cost-Plus*
- *EJCDC D-610, Design-Build Contract Performance Bond*
- *EJCDC D-615, Design-Build Contract Payment Bond*
- *EJCDC D-700, Standard General Conditions of the Contract Between Owner and Design-Builder*
- *EJCDC D-750, Standard General Conditions of the Subcontract Between Design-Builder and Subcontractor*

DBIA has developed documents guiding the owner and design-builder in the design-build process. Various documents include:

- *DBIA Document 520, Standard Form of Preliminary Agreement Between Owner and Design-Builder*
- *DBIA Document 525, Standard Form of Agreement Between Owner and Design-Builder—Lump Sum*
- *DBIA Document 530, Standard Form of Agreement Between Owner and Design-Builder—Cost Plus Fee with an Option for a Guaranteed Maximum Price*
- *DBIA Document 535, Standard Form of General Conditions of Contract Between Owner and Design-Builder*
- *DBIA Document 540, Standard Form of Agreement Between Design-Builder and Designer*
- *DBIA Document 550, Standard Form of Agreement Between Design-Builder and General Contractor—Cost Plus Fee with an Option for a Guaranteed Maximum Price*
- *DBIA Document 555, Standard Form of Agreement Between Design-Builder and General Contractor—Lump Sum*
- *DBIA Document 560, Standard Form of Agreement Between Design-Builder and Design-Build Subcontractor—Cost plus Fee with an Option for a Guaranteed Maximum Price*
- *DBIA Document 565, Standard Form of Agreement Between Design-Builder and Design-Build Subcontractor—Lump Sum*

The AGC forms of design-build contracts recognize preliminary and final phases. AGC publishes the following contract documents:

- *AGC ConsensusDOCS 410, Standard Form of Design-Build Agreement and General Conditions Between Owner and Design-Builder* (Where the Basis of Payment Is Cost of the Work Plus a Fee with GMP)
- *AGC ConsensusDOCS 415, Standard Form of Design-Build Agreement and General Conditions Between Owner and Design-Builder* (Where the Basis of Payment Is a Lump Sum Based on an Owner's Program Including Schematic Design Documents)
- *AGC ConsensusDOCS 420, Standard Form of Agreement Between Design-Builder and Architect/Engineer for Design-Build Projects*
- *AGC ConsensusDOCS 421, Design-Builder Statement of Qualifications for a Specific Project*
- There are also several agreements with subcontractors as well as performance and payment bonds in this 400 series

Division 01—General Requirements The design-builder should prepare the general requirements as part of the construction documents. The Division 01 sections for a design-build project delivery are developed for the construction and might involve a contractor or various subcontractors. Division 01 may also be used to assign temporary construction requirements to various subcontractors. Some of the sections included in Division 01 as they apply to design-build contracts are summarized in the following paragraphs.

- *Section 01 10 00—Summary.* Should identify responsibilities and owner-furnished items, partial occupancy, and use of the site.
- *Section 01 21 00—Allowances.* May be used to delay design decisions. Allowances permit the design-builder to achieve accurate cost control even if some products are not known. These unknowns should be minimized because one of the factors in selecting the design-build delivery method is that the project requirements are reasonably predictable.
- *Section 01 25 00—Substitution Procedures.* Should be included in the design-build construction documents. There are always circumstances that require substitutions. The substitution procedures should be managed so that the owner is aware of the substitutions and products ultimately included in the project.
- *Section 01 33 00—Submittal Procedures.* Should be included for consistency in the information required for submittals. In a design-build contract, the design-builder is usually required to maintain a complete set of submittals at the project site. The submittal procedure in the design build process is the same as in the design-bid-build delivery method. Shop drawings, samples, and product data are received and checked by the design-builder and distributed by the design-builder or returned to the subcontractor for resubmittal. The owner should also receive a set of approved shop drawings.
- *Section 01 50 00—Temporary Facilities and Controls.* Includes requirements for the use of parking and staging areas, storage sheds, trailers, hoists, sanitary facilities, fire protection, and temporary utilities. Fences and other barriers, walkways, project signs, and minor temporary construction items are also specified in this section. Assignment of temporary construction to an appropriate subcontractor may be specified in this section. It is usually necessary to include specification requirements for temporary facilities and controls, even if the design-builder will be performing the work. The owner may want to see this information as part of the contract documents.
- *Section 01 70 00—Execution and Closeout Requirements.* Should be included for the orderly closing out of the design-builder's and subcontractors' duties, such as submission of warranties, maintenance data, maintenance materials, project record documents, and provisions for final cleaning. The design-builder inspects the project for substantial completion. The results of this inspection are presented to the owner for evaluation.

Divisions 02 through 49—Specifications The design-builder should prepare specifications to secure accurate subcontract bids and ensure clear definition of the responsibilities of various subcontractors. The level of detail of the specifications may be influenced by the design-builder or by the owner's project description. The design-builder may desire brief specifications that allow wide latitude in product requirements and selections, or the owner may have stringent project requirements that dictate the level of specification detail required.

1.14.4 Document Variations for Owner-Build Project Delivery Method

Construction documents prepared for the owner are those required by the nature of the project extent. A project may range from simple maintenance or rehabilitation to a complete new facility. The documents are structured similar to those of other delivery methods. A simple project may bid the same as a traditional design-bid-build delivery method or may be separated into multiple contracts similar to construction management delivery methods. An owner might even consider a simple design-build delivery method for a pre-engineered structure. Owners who are capable of preparing their own designs and documentation might have an in-house department with an A/E or facility manager. Documentation may be performed in-house or under agreement with an independent A/E.

Project Manual Variations Project manual preparation for owner-build projects is similar to that for other project delivery methods. A project may require a single project manual or several project manuals for multiple-prime contracts.

Division 00-Procurement and Contracting Requirements

Procurement Requirements The procurement requirements will be developed similar to other delivery methods and will depend on the nature of the project and whether it is to be bid, negotiated, purchased, or some form of services. The owner might prepare the owner's own procurement requirements for subcontract packages with specialty contractors.

Contracting Forms The owner-contractor agreement included in a project manual might be a standard form or determined by informal negotiations. The A/E might not always participate in this process. The A/E might simply include the owner's standard documents that have been prepared by the owner's attorney or in-house legal department. A copy of the owner-contractor agreement form should be available to ensure there are no conflicts between the agreement and the conditions of the contract or Division 01. If the A/E is in independent practice, the A/E should review these forms to ensure they do not contain responsibilities not included in the A/E agreement. The A/E should also inform the owner if the proposed agreement does not require responses to the A/E's responsibilities.

Conditions of the Contract The conditions of the contract may also be standard documents that have been prepared by the owner's attorney or in-house legal department. The conditions of the contract should appropriately reflect the primary aspect of the similar project delivery methods. This means that provision might be required to ensure coordination between multiple contracts or specific issues for purchases or services. The conditions of the contract affect the various participants involved. The conditions of the contract are the basis of the requirements established in Division 01. Conditions of the contract should be included in the project manual to provide continuity of these requirements.

If the conditions of the contract are not industry standard, such as AIA or EJCDC General Conditions, the A/E must carefully review the documents to understand their variations from the standard documents. Variations in general conditions might involve A/E and contractor roles different from those normally associated with standard agreements. These requirements or responsibilities must be reconciled.

Requirements for payments and changes may be covered in the owner-contractor agreement and should not be repeated in Division 01. If the owner-contractor agreement is not available to the A/E, the A/E should determine appropriate circumstances for preparation of the project manual.

Division 01—General Requirements Possible conflicts among the agreement, the conditions of the contract, and the Division 01 requirements can occur, depending on their source and particular project requirements. Division 01 requirements should be prepared in conjunction with the conditions of the contract. Specific issues should address operations at existing facilities including use of the site, security, personnel identification, and access to the work. Other issues may involve certain procedures such as payment procedures, construction progress schedules, submittals, substitutions, warranties, and contract closeout.

Divisions 02 through 49—Specifications Writing specifications for this delivery method are not usually different; however, some firms will have specific requirements for products that have become standard in their identity or operations. The decisions to utilize specific products might require the owner's specific approval. If the owner requires specific products, then the specification might be written as proprietary, eliminating a level of detail.

Chapter 2
Language

2.1 Specification Language

As legally enforceable contract documents, construction specifications should be prepared with concern and respect for their legal status. The four Cs for effective communication are:

Clear Use proper grammar and simple sentence construction to avoid ambiguity.

Concise Eliminate unnecessary words, but not at the expense of clarity, correctness, or completeness.

Correct Present information accurately and precisely. Carefully select words that convey exact meanings.

Complete Do not leave out important information.

2.2 Writing Style

Good writing style is characterized by accuracy, brevity, and clarity. Long, complex sentences and stilted language do not contribute to effective communication. Follow these simple rules for good style:

- Use simple sentences.
- Maintain sentence structure in simple declarative or imperative statements.
- Avoid complicated sentences in which inadvertent omission or insertion of punctuation could change meaning or create ambiguity.
- Choose words and terms that are simple and clearly understood.

2.3 Vocabulary

Words should be carefully selected and used for precise meaning. Once a word is selected, use it consistently throughout the specifications whenever the same meaning is intended. The following are some examples of commonly misused or ambiguous terms, along with guidelines for their recommended usage in specifications.

Amount and Quantity *Amount* should be used when writing about money. *Quantity* should be used when writing about number, linear measure, area, or volume.

Any *Any* is imprecise in number, permitting discretion by the reader. "Repair any cracks" could mean some cracks selected by the contractor, but "repair cracks" means all cracks.

And, Or, And/Or *And* connects elements that are to be taken jointly. It may also mean "plus" or "added to the preceding quantity." *Or* is used to introduce any of the possibilities in a series. The two words together, *and/or,* represent a hybrid term often used in legal and business documents as a grammatical shortcut. The term *and/or* is not recommended for specifications, because it allows the contractor to decide whether the term means *and* or *or.*

Balance and Remainder *Balance* should be used when writing about money. *Remainder* is preferred for "that which is left over."

Either and Both *Either* implies a choice between two options, whereas *both* is all inclusive. Make clear whether the intent is to have "glass sidelights on either side of the door" or to have "glass sidelights on both sides of the door."

Flammable and Inflammable *Flammable* and *inflammable* have exactly the same meaning, even though they sound opposite. *Flammable* is preferred.

Furnish, Install, and Provide *Furnish* means to supply and deliver to the project site, ready for installation. *Install* means to place in position for service or use. *Provide* is commonly accepted in specifications to mean furnish and install, complete and ready for intended use. These definitions should be placed in Supplementary Conditions or in Division 01—General Requirements.

Insure, Assure, and Ensure To *insure* is to issue or procure an insurance policy. *Assure* is to give confidence to or convince a person of something. *Ensure* is to make certain in a way that eliminates the possibility of error.

Observe and Supervise *Observe* means to watch or view the execution or performance of work, whereas *supervise* means to oversee and to have control and direction of the work.

Party and Entity Party refers to a signer of a contract, such as the owner and contractor in an owner-contractor agreement. When the intent is to include persons or companies, such as subcontractors and others who are involved in the construction process but are not signers of the contract, the generic *entity* should be used.

Replace and Provide New *Replace* can be construed as requiring reinstallation of removed materials rather than furnishing and installing new materials. Be specific as to what is required.

Shall and Will *Shall* is used as an imperative in reference to the work required to be done by a contractor. *Will* is optional and is used in connection with acts and actions required of the owner or the architect/engineer (A/E).

Must and *is to* are not recommended.

2.4 Spelling

Spelling should be correct and consistent, based on a particular dictionary designated as the office standard for spelling. A supplemental list of technical terms may be needed for words not contained in the dictionary selected as the standard. In cases where two spellings are considered equally correct, the shorter of the two spellings is preferred for use in contract documents (e.g., *calk* not *caulk, facia* not *fascia, gage* not *gauge, molding* not *moulding,* and *catalog* not *catalogue*). However, unacceptable spellings produced by attempts at brevity or simplification should be avoided (e.g., *thru* instead of *through*).

2.5 Sentence Structure

Two basic grammatical sentence moods can be used to clearly convey specification requirements:

- Imperative mood
- Indicative mood

2.5.1 Imperative Mood

The imperative mood is the recommended method for instructions covering the installation of products and equipment. The verb that clearly defines the action becomes the first word in the sentence. The imperative sentence is concise and readily understandable:

- Spread adhesive with notched trowel.
- Install equipment plumb and level.
- Apply two coats of paint to each exposed surface.

2.5.2 Indicative Mood

The indicative mood, passive voice requires the use of *shall* in nearly every statement. This sentence structure can cause unnecessary wordiness and monotony:

- Adhesive shall be spread with notched trowel.
- Equipment shall be installed plumb and level.
- Two coats of paint shall be applied to each exposed surface.

2.6 Abbreviations

Abbreviations should be used only on drawings and schedules where space is limited. Well-known and industry-accepted abbreviations are a type of shorthand that helps the communication process only if the meaning is easily recognized and understood. Abbreviations should be defined on the drawings or in Division 01. Abbreviations with multiple meanings should be avoided, unless used in different disciplines where their meaning is clear from the context in which they are used. Abbreviations should be limited to five or fewer letters. Abbreviations of short words that save only one or two characters should usually be avoided. A list of abbreviations is included in the

U.S. National CAD Standard for Architecture, Engineering, and Construction (NCS). When in doubt, spell it out.

2.7 Symbols

The cautions and guidelines for abbreviations apply to symbols substituted for words or terms. Two additional factors that limit the use of symbols are their conflict with use as command characters in software programs and potential translation of font problems when converting from one software program to another. Small symbols may also bleed together and become unreadable in a poorly printed text. Following are some of the symbols that should not be used in specifications:

- % for percent
- ° for degree
- + for plus
- – for minus
- • for by, as in 2 × 4
- / for per
- @ for at

The use of parentheses and quotation marks should be minimized or avoided; underlines should not be used. When including dimensions in the text of a specification section, apply the following rules:

- Spell out feet when no inches are used (e.g., 8 feet).
- Spell out inches when no feet are used (e.g., 8 inches).
- When feet and inches are both used, use symbols (e.g., 8′-8″or 8′ -2–1/2).
- A complete dimension should appear on one line.

Standards designations and *MasterFormat*® number such as ASTM C 270, ANSI/ ASME A17.1, ACI 318, and Section 01 42 19 should appear on the same line; do not separate them.

2.8 Numbers

The use of Arabic numerals rather than words for numbers is recommended in accordance with the following rules:

- Use numerals whenever possible because they are easy to identify. However, when numbers are used to define both size and quantity, the written word should be used for the quantity (e.g., three 1/2-inch holes; five 2 by 4s).
- Always use figures for dimensions, degrees of temperature, percentages, and dollars and cents (e.g., 3 inches by 5 inches, 10 degrees C (50 degrees F), 20 percent, $5.50).

- Clock times and dates should be expressed in figures (e.g., 2:10 P.M. on June 15 (omit "th"), 2007. Exceptions to this are the use of the words *noon* and *midnight.* The same rules apply to a 24-hour clock as well.
- Decimals should be expressed in figures (e.g., 6.235). For quantities less than one, a zero should be used before the decimal point (e.g., 0.235).
- To maintain uniformity throughout the specification, fractions should be typed using individual keys and not converted to superscript/subscript fonts (e.g., 1/4 not 1/4 and 1/2 not 1/2).
- Omit unneeded zeroes in time and money references (e.g., $200 not $200.00 and 9 P.M. not 9:00 P.M.).

2.9 Capitalization

Capitalization should be consistent throughout the contract documents. Capitalization of the initial letter of certain specific nouns and of proper names defined in the conditions of the contract is appropriate. Following are some examples of words that should be capitalized:

Agreement When referring to the specific form signed to execute the contract.

Architect When referring to the architect who is a party to the owner-architect agreement.

Article When referring to an article in the specifications or conditions of the contract.

Change Order When issued as a modification to a contract.

Contract When referring to the specific contract for which the specifications are written.

Contracting Officer When referring to the representative of a government agency with authority to make decisions on behalf of the agency.

Contractor When referring to the contractor who is party to the owner-contractor agreement.

Division When referring to a specific division within the project manual.

Drawings When referring to graphic portions of the contract documents.

Engineer When referring to the engineer who is a party to the owner-engineer agreement.

General Conditions When referring to the specific general conditions of a contract.

Government When a government agency is a party to the contract.

Owner When referring to the owner who is a party to the owner-contractor or owner-architect agreement.

Paragraph When referring to a paragraph in the specifications or other contract documents.

PART When referring to one of the parts of *SectionFormat*™.

Project When referring to the specific project of which the work is a part.

Project Manual When referring to the bound volume that is part of a specific set of construction documents to be defined in the Supplementary Conditions.

Room Names For example, Library, Science Room, or Chemistry Laboratory.

Section When referring to a specific section of the specifications.

Shop Drawings, Product Data, and Samples When referring to submittals required for the specific project.

Specifications When referring to sections in Divisions 01–49.

State or Commonwealth When referring to a specific state (e.g., the State of Missouri, the Commonwealth of Pennsylvania).

Supplementary Conditions When referring to the specific modifications to the general conditions of a contract.

Work When referring to the work of a specific contract or a portion of the project.

No capitalization is required when the preceding examples are used in the general sense. Directions such as east or northwest are not capitalized unless they form a proper noun. The words *subcontract* and *subcontractor* are not capitalized because they do not apply to a specific party defined in the contract agreement. The words *subparagraph* and *clause* are usually not capitalized.

2.10 Punctuation

Because specifications are legal documents, the formal rules of punctuation must be observed. Sentences should be constructed so that the misplacement or elimination of a punctuation mark will not change the meaning. Commas should be used after each item in a series, including the item preceding a conjunction, and in other locations where the clarity of the statement will be improved.

2.11 Grammar

2.11.1 Subject/Verb Agreement

The subject and the verb must always agree in number. Singular verbs should be used with singular subjects and plural verbs with plural subjects. An error in number is easy to make when a sentence is long and complicated. The singular subject of a sentence can be confused with a plural modifier.

Incorrect One of the elongated central fasteners are to be placed around the eye of the panel and bolted.

Correct One of the elongated central fasteners shall be placed around the eye of the panel and bolted.

Preferred Bolt one elongated central fastener to panel eye.

The incorrect example uses the singular subject *one* with the plural verb *are.* The grammatically correct example has number agreement between subject and verb, but is an involved and relatively verbose sentence. The preferred language has number agreement and is a simple, direct statement of instruction—that is, clear, concise, correct, and complete.

2.11.2 Parallel Construction

Good grammar also requires the use of identical style in both parts of a compound subject or predicate. The use of identical style in a series of nouns, adverbs, or prepositional phrases is also recommended.

Incorrect Tests shall be performed to determine strength and establish qualities.

Correct Tests shall be performed to determine strength and to establish quality.

Preferred Perform tests to determine strength and to establish quality.

Incorrect Heating, ventilation, and air-conditioning.

Correct and Preferred Heating, ventilating, and air-conditioning.

2.12 Inappropriate Terms

Avoid using phrases that have missing objects:

- As allowed
- As appropriate
- As approved
- As directed
- As indicated
- As necessary
- As required

Avoid these adverbs:

- Hereinafter
- Hereinbefore
- Herewith
- Wherein

Avoid these articles:

- Any
- All
- Such

Avoid these words and expressions:

- Etc.
- As per
- In a workmanlike manner
- To the satisfaction of the architect/engineer
- Shall function as intended
- Also

2.13 Pronoun Reference

The use of pronouns in specifications should be minimized or avoided. Personal pronouns should not be used. Repeating the noun is better than risking possible misunderstanding.

Poor Apply coating with pneumatic equipment when it is above 40°F.

Better Maintain pneumatic equipment above 5 degrees C (40 degrees F) or apply coating only when ambient temperature is above 5 degrees C (40 degrees F).

Which and other relative pronouns should be used sparingly, if at all.

Poor Contractor shall install bathroom accessories which are to be purchased under an allowance.

Better Contractor shall install bathroom accessories to be purchased under an allowance.

Preferred Install bathroom accessories purchased under allowances specified in Section 01 21 00.

The word *same* should not be used as a pronoun.

Poor If materials are rejected, the contractor shall replace same at no additional cost.

Better Contractor shall replace rejected materials.

Preferred Replace rejected materials.

2.14 Unnecessary Words

Definite article *the* and indefinite articles *a* and *an* need not be used in most instances.

Poor Apply an oil paint with a brush to the wall.

Better Apply oil paint with brush to walls.

The use of the word *all* is usually unnecessary.

Poor Store all millwork under shelter.

Better Store millwork under shelter.

Avoid using *contractor* as the subject of the sentence.

Poor Contractor shall lay brick in common bond.

Better Brick shall be laid in common bond.

Preferred Lay brick in common bond.

2.15 Prepositional Phrases

Sentences may be shortened in specification language by using modifiers in place of prepositional phrases.

Correct Top of platform.
Preferred Platform top.
Correct Within the time recommended by manufacturer.
Preferred Within manufacturer's recommended time limit.

2.16 Streamlining

Attempts to reduce verbiage in specifications are recommended if the meaning can still be clearly conveyed. Although difficult to adapt to descriptions or instruction, streamlining is used to list products, materials, reference standards, and other itemized specifications. This technique places the subject first and provides keywords for quick reference.

- Adhesive: Spread with notched trowel
- Equipment: Install plumb and level
- Portland cement: ASTM C 150, Type 1
- Aggregate: ASTM C 33
- Air-entraining agent: More-Air Brand, More-X Manufacturing Company.

When using the streamlining method, A/Es often include an explanatory statement in Division 01 or the Supplementary Conditions:

> These specifications are written in imperative mood and streamlined form. This imperative language is directed to the Contractor, unless specifically noted otherwise. The words "shall be" are included by inference where a colon (:) is used within sentences or phrases.

2.17 Specification Detail

Language style should not be confused with specification detail. Specification detail should be commensurate with the requirements of the project and method of project delivery. Specifications for a large housing project may be more complex than those for a small vacation cottage, but the same general rules for clarity and conciseness apply to both projects.

Degree of detail is a matter of judgment and is often tempered by economic considerations. A specification is complete when it covers important details without elaborate or unnecessary language.

Chapter 3
Methods

3.1 Methods of Specifying

There are four methods of specifying: descriptive, performance, reference standard, and proprietary.

- *Descriptive.* Specifies properties of materials and methods of installation without using proprietary names. A descriptive specification is a detailed description of the characteristics, physical properties, and workmanship required for the installation of a product or material. It generally requires technical knowledge and experience on the part of the architect/engineer (A/E).
- *Performance.* Specifies the required results, the criteria by which the performance will be judged, and the method by which it can be verified. The contractor is free to choose materials and methods complying with the performance criteria. A performance specification is a description of the required end result of a product or system and includes the criteria to be used for verification of proper installation. It is generally written to encourage the use of innovative techniques.
- *Reference Standard.* Specifies product or processes by established standards. A reference standard specification uses recognized industry standards rather than individually written product or installation criteria. Standards must be reviewed carefully to avoid duplications, optional requirements, and contradictions.
- *Proprietary.* Specifies actual brand names, model numbers, and other proprietary information. A proprietary specification describes products and materials by manufacturer's name, brand name, model number, or unique characteristic. Closed proprietary specifications do not allow for substitutions, but open proprietary specifications allow for alternative products.

Both the descriptive and proprietary specification methods are prescriptive; that is, the products and processes are specified, but not the results. Reference standards are used for prescriptive or performance specifications. Performance, however, specifies results and not the means to achieve them. Project specifications typically employ more than one specifying method. All four methods may be used in a single specification section. There is no clear rule for using either one method or a combination of methods. However, the architect/engineer (A/E) should be careful about combining methods in the specification of a single product. This approach can create redundancy or may result in a conflict incapable of resolution.

3.1.1 Descriptive Specifications

A descriptive specification is a detailed, written description of the required properties of a product, material, or piece of equipment and the workmanship required for its installation. Proprietary names of manufacturers are not used. A concrete mix of four-part coarse aggregate, two-part fine aggregate, and one-part cement with a 0.5 water-cement ratio is a descriptive specification. Performance strength of 20,684 kPa (3000 psi) after 28 days is implied in the design mix, but not specified. If concrete conforming to the descriptive specification were supplied but did not withstand a 20,684 kPa (3000 psi) load, the contractor could not be held responsible because only the design mix was specified. The burden of performance is assumed by the A/E when a descriptive specification is used. Once widely preferred, the descriptive method is being used less frequently as projects become more complex and as better reference standards become available. Writing a descriptive specification is a lengthy and tedious process. However, when proprietary names are prohibited by law and when adequate reference standards do not exist, a descriptive specification may be the only logical choice.

3.1.1.1 Preparing Descriptive Specifications

There are five basic steps for preparing descriptive specifications:

1. Research available products.
2. Research important features required. Analyze and compare requirements with available products.
3. Determine which features are best specified and which are best shown on the drawings.
4. Describe important features. State the requirements and ascertain that they can be met. Selecting and specifying unique features from the products of several manufacturers may result in the description of a nonexistent product. Avoid specifying unnecessary features.
5. Specify only information about submittals, testing, and other procedures necessary to ensure that acceptable products will be provided. Do not require extensive shop drawings if catalog information is adequate. Do not require tests if product certification will suffice.

Figure 3.1 shows a portion of a descriptive specification for a shelf support assembly. Information on location, length, and spacing of the supports would be shown on the drawings.

3.1.2 Performance Specifications

A requirement of 20,684 kPa (3000 psi) concrete strength is a performance specification. The end result, rather than the means to the end result, is specified. A performance specification is defined as a statement of required results with criteria for verifying compliance, but without unnecessary limitations on the methods for achieving the required results.

"A statement of required results" means that desired end results must be spelled out. An incomplete performance specification results in a major loss of quality control over the materials, equipment, and workmanship going into a project.

PART 2—PRODUCTS
2.01 SHELF STANDARDS

A. **Standards:**
1. Surface-mounted, projecting 25 mm (1 inch) maximum from finished surface.
2. Mounting screws spaced 300 mm (12 inches) apart maximum.
3. Slots for shelf adjustment on approximate 25 mm (1 inch) centers.
4. Finish: Gold anodized aluminum.

B. **Mounting Screws:**
1. Size: No. 8 by 40 mm (1-1/2 inch) long.
2. Type: Steel wood screws.
3. Finish: Match standards.

C. **Shelf Brackets:**
1. Size: Suitable for 200 mm (8 inches) wide shelves.
2. Depth at butt: 75 mm (3 inches) maximum.
3. Depth at tip: 25 mm (1 inch) maximum.
4. Attachment: Rigid engagement of lugs into 2 slots of standard.

Figure 3.1
Sample of a Descriptive Specification

"With criteria for verifying compliance" means that the criteria are capable of measurement, test, evaluation, or other acceptable assurances. Measurement and testing may be done before production, at the time of production, in place at the site, or after a period of service.

"Without unnecessary limitations on the methods for achieving the required results" means that only essential restrictions are placed on the system. Limitations on the means should be avoided. Performance specifying should keep specific material and process descriptions to a minimum to encourage the devising of new means to achieve desired results.

3.1.2.1 Application of Performance Specifying

The term *performance specifying* as related to project specifications has two applications. First, simple performance criteria can be incorporated into any specification, even one that generally describes what is wanted. A ceiling hanger may be performance specified as "non-sag" rather than descriptively as "straightened No. 11 wire, tightly wrapped two times."

In its second and broader application, performance specifying is a way of obtaining improved products and methods by stating the desired result and leaving the means to the innovative producer or contractor. What is often specified in this case is a systems approach.

Where performance specifying is the primary mode of design and contracting procedure, specialized contract documents are often required. Contract procedures are more complex and often involve a variety of participants.

It is possible to successfully combine performance specifying and descriptive specifying in the same project. Some products may be specified with descriptions of components, and other products with performance statements. However, using both descriptive and performance modes for a single requirement should be avoided because it would be redundant and open to contradictions.

If it is known how a specific item performs, it is not difficult to work backward and describe its performance in sufficient detail so that bidders or proposers will know exactly which product is desired. In such cases, the writing of the specification takes much more time and effort and secures nothing better than what would result from descriptive or proprietary specifying. It is much easier to ask for a 10-gage galvanized, annealed steel wire than to specify each performance criterion.

It is crucial that the performance requirements stipulated can be met. To establish criteria that no manufacturer can meet would result in unachievable requirements. For a detailed description on performance specifications, refer to Chapter 12, "Performance Specifying."

3.1.2.2 Decision to Use Performance Specifying

Early in the planning process, the owner and the A/E should determine the suitability of the project to the performance specifying concept. The owner's needs may make performance specifications feasible for several reasons:

- To expedite construction and access a wide range of options using existing technology and systems techniques
- To utilize technology that has not yet become standardized within the construction industry
- To develop new technology
- To delegate technical design responsibilities to industry specialists

3.1.3 Reference Standard Specifications

A standard is a document established by consensus that provides rules, guidelines, or characteristics for activities or their results as defined in the International Organization for Standardization/International Electrotechnical Commission (ISO/IEC) Guide 2. Standards are incorporated by reference into the specifications and thus become commonly known as reference standards. They are published by trade associations, professional societies, standards-writing organizations, governments, and institutional organizations. Typical authors are architects, engineers, scientists, technologists, manufacturers, and product users who are extremely knowledgeable about the particular reference standard subject. Where applicable, use standards listed in the model or local codes to ensure acceptability with the authority having jurisdiction (AHJ). Depending on how it is listed in the codes, such a standard may be considered either a code requirement or a guideline for acceptance by the AHJ.

Reference standards include the following types:

- Basic materials standards such as American Society for Testing and Materials (ASTM) B 211, Standard Specifications for Aluminum and Aluminum-Alloy Bar, Rod, and Wire
- Product standards such as Engineered Wood Association (APA) APA PS-1, Construction and Industrial Plywood
- Design standards such as American Concrete Institute International (ACI) ACI 318, Building Code Requirements for Structural Concrete
- Workmanship standards such as ASTM E 2112, Standard Practice for Installation of Exterior Windows, Doors and Skylights

- Test method standards such as ASTM E 119, Standard Test Methods for Fire Tests of Building Construction and Materials
- Codes such as American National Standards Institute/American Society of Mechanical Engineers (ANSI/ASME) A17.1, Safety Code for Elevators and Escalators
- Installation standards such as National Fire Protection Association International (NFPA) 13, Standard for the Installation of Sprinkler Systems
- Performance standards such as ASTM F 1487, Standard Consumer Safety Performance Specification for Playground Equipment for Public Use
- Life safety codes and standards produced by organizations such as International Code Council, Inc. (ICC), NFPA, and Underwriters Laboratories, Inc. (UL)

Reference standards are incorporated into the specifications by referring to a number, title, or other designation. The provisions of standards so referenced become a part of the specifications just as though included in their entirety. Their incorporation into the specifications by reference saves the A/E the work of writing an elaborate and lengthy text. Accompanying these benefits are some liabilities:

- Inadequate reference standards coexist with stringent ones.
- Reference standards can create duplication and contradiction within the contract documents.
- Standards can contain embedded options.
- Standards generally refer to minimum requirements.
- Reference standards might contain undesired requirements.
- Various AHJs may enforce different editions of the same standard, which might have conflicting requirements.

Adherence to the following three guidelines will help eliminate problems associated with reference standards:

- Know the standard.
- Incorporate the standard properly.
- Enforce the requirements of the standard.

3.1.3.1 Know the Standard

A reference standard should be reviewed and its content and purpose thoroughly understood before it is incorporated into the specifications. Other project requirements must be compatible with the reference standard, and the standard should be free of duplications and contradictions. If the reference standard does not meet these criteria, the A/E will have to choose between using the standard as a reference and defining exceptions to it or writing original requirements into the specifications.

The A/E should have access to the reference standards cited in the specifications. Maintaining a current library of all reference standards is an awesome task, but it is advisable to obtain commonly referenced standards.

3.1.3.1.1 Duplications and Conflicts Duplications and conflicts may occur when two or more reference standards are used for a given product. It is possible that the requirements of one standard may conflict with another, thus creating an opportunity for the use of a product conforming to lesser requirements than those that were intended. Some reference standards contain provisions that, at first glance, appear to be an irrevocable part of the standard, but a footnote or subsequent paragraph may void these provisions unless they are specifically stated in the specifications.

Reference standards, especially those dealing with workmanship, may contain statements similar to those in the conditions of the contract or Division 01 of the specifications. When standards are referenced, the duplicated statements become part of the project specifications. Statements that differ in wording or intent from requirements of the conditions of the contract or Division 01 can create ambiguities leading to conflicts. For example, the American Institute of Steel Construction, Inc. (AISC) Code of Standard Practice covers numerous contractor-owner and contractor-subcontractor relationships. The AISC Code also uses the words *owner* and *contract documents* in a different context than that in the widely used general conditions of a construction contract such as *American Institute of Architects (AIA) Document A201, Conditions of the Contract for Construction*, and *Engineers Joint Contract Documents Committee (EJCDC) C-700, Standard General Conditions of the Construction Contract.* These situations could easily contribute to misunderstandings.

In addition, other provisions often appear in reference standards, such as the following examples:

- "Spaces above suspended ceilings shall be ventilated."
- "During the installation of drywall tape, the building temperature shall be maintained at not less than 10 degrees C (50 degrees F)."

Statements such as these might conflict with other requirements in the specifications and can create confusion, misinterpretation, or contradictions. Wherever the wording or intent of such provisions differs, there exists a potential for disputes. Such provisions should be identified and clearly excluded when the standard is used. Similarly, the A/E could state that when there is a conflict or discrepancy between a reference standard and the specifications or another referenced standard, the more stringent requirements shall apply. Such statements cannot be relied upon to handle all possible conflicts, however. It is important for the A/E to identify the duplications and modify the references to eliminate contradictions.

3.1.3.1.2 Embedded Options Many standards include several categories, classes, or groups from which applicable properties must be selected. These embedded options constitute choices that must be identified and modified to prevent discrepancies. For example, ASTM C 150/C 150M, Standard Specification for Portland Cement, covers 10 types of cement. The A/E should evaluate each choice. If the A/E does not specify the type, the selection is forfeited to the contractor, who can reasonably be expected to select the least expensive option.

3.1.3.1.3 Multiple Standards Multiple standards are those containing references to other standards. They create conflicts of information and produce multiple choices in the same manner as single reference standards. The AISC Document "Specifications for the Design, Fabrication, and Erection of Structural Steel for Buildings" provides another example. This document refers to several ASTM standards, and these refer, in turn, to still other ASTM standards. The essential point is that standards listed within other standards must not contain qualifications that will contradict requirements of any other referenced standard or the project specification. There is no easy solution to this problem other than to know and understand the content of the additional referenced material.

3.1.3.1.4 Trades and Subcontracts Reference standards may refer to a particular trade or to a particular subcontractor. Construction contracts are usually written for one contractor, and reference to a particular subcontractor in a project specification should be avoided. Again, knowing the contents of the reference standard will permit elimination of statements that contradict the requirements of the project specifications.

3.1.3.1.5 Level of Requirements Reference standards often define minimum requirements. The minimum requirements may be so restrictive that they exclude most commercially available materials or may be so liberal that nearly anything produced can meet them. To avoid these extremes, exceptions may be made to the reference standards.

3.1.3.2 Incorporate the Standard Properly

Incorporating the standard properly requires the use of designations that distinguish each standard from all of the others. A complete designation includes:

- Name of the issuing organization
- Number of the standard
- Title of the standard
- Date of issue of the standard
- Citation of applicable requirements unless the entire standard is required

The issuing organization is often referenced by its initials or abbreviation. Many organizations and their abbreviations are well known; others are not so widely recognized. It is appropriate to spell out the organization's name along with its initials or abbreviation in parentheses when it is first mentioned in the specification section (refer to Figure 3.2, Option 1). A second option is to define all abbreviations of reference standards organizations

"OPTION 1"
1.05 REFERENCES

A. American Concrete Institute International (ACI):
   1. 301-99 Specifications for Structural Concrete for Buildings.
   2. 304R-00 Guide for Measuring, Mixing, Transporting and Placing Concrete.

B. American Society for Testing and Materials International (ASTM):
   1. C33-03 Standard Specification for Concrete Aggregates.
   2. C150-02a Standard Specification for Portland Cement.

"OPTION 2"
1.05 REFERENCES

A. ACI 301 Specifications for Structural Concrete for Buildings.
B. ACI 304R Guide for Measuring, Mixing, Transporting and Placing Concrete.
C. ASTM C 33 Standard Specification for Concrete Aggregates.
D. ASTM C 150 Standard Specification for Portland Cement.

NOTE: It is recommended that if "Option 2" is used, a general statement in Division 01—General Requirements should be included to establish the dates of the standards referenced.

Figure 3.2
Alternative Methods for Incorporating Reference Standards

in Division 01, Section 01 42 00—References, as shown in Figure 3.2, Option 2. Section 01 42 00 may also include the addresses of the organizations whose standards are referenced in the project manual. The reference standard should be designated in exactly the same manner as it is printed on the actual standard. When the standard is referenced in a specification section, its complete designation should be included in the article "References" in PART 1—GENERAL as described in *SectionFormat™*. Then, subsequent references to the standard in PART 2—PRODUCTS and PART 3—EXECUTION are referenced only by the number. The complete listing of standards referred to in PART 2 and PART 3 under the "References" article facilitates the gathering of reference materials. This listing of references can contain addresses to web sites where the standards are available for view or purchase. The shorter references in PARTS 2 and 3 save time, effort, and space. The standards shown in Figure 3.2 would be referenced in PARTS 2 and 3 of the specification section as ACI 301, ACI 304R, ASTM C 33/C 33M, and ASTM C 150/C 150M.

3.1.3.2.1 Use of Dates Standards are revised periodically. The revisions are dated with the year and sometimes the month of issue. As shown in Figure 3.2, Option 1, for ASTM C 33/C 33M-08 and ASTM C 150/C 150M-09, the 08 and 09 indicate that the standards were revised and republished in 2008 and 2009, respectively. There are two approaches to setting the dates of reference standards. One is to include the dates of reference standards (refer to Figure 3.2, Option 1). By including the dates, the A/E implies that the edition of the standard referenced fulfills the requirements of the project. It is the A/E's responsibility to determine the date of the applicable or latest edition. This task is made easier because standards organizations typically publish periodic catalogs. The A/E should not merely alter a publication date without reviewing the changes that have been made to the standard. A possible problem with stating dates is that standards current when the specification is prepared may be revised by the time the project is bid or construction begins. AHJs might not have adopted or enforced the latest edition of a standard.

The second approach (refer to Figure 3.2, Option 2) is not to list publication dates for reference standards. This action alone is not recommended, as numerous editions of the same standard could result in contradictions. A solution to this problem is to use the following statement in Division 01, Section 01 42 00—References: "The date of the standard is that in effect as of the date of receipt of bids for the project." With such a statement, the contractor becomes responsible for checking for compliance (refer to Figure 3.2, Option 2).

3.1.3.2.2 Tentative Draft Standards Standards listed as tentative have been approved by the sponsoring organization but are awaiting final approval by its membership or by other concerned organizations. Tentative draft standards usually represent the latest thoughts and practices of the sponsoring organization. Tentative draft standards are usually not accepted by AHJs. A/Es should use tentative standards with caution because of their unofficial status.

3.1.3.2.3 Quoting Selected Parts of Standards The practice of referencing a standard and then quoting supposedly pertinent portions of it should be avoided. The reader may believe that only the quoted parts apply. In a similar manner, a specific clause in a standard should not be referred to unless the requirements of that clause are being modified. Information may be incorrectly quoted.

3.1.3.3 Enforce the Requirements of the Standard

Once specified, effective means must be provided to ensure conformance with the requirements of the standards. Usual methods include checking shop drawings, samples, manufacturer's literature, test reports, and other required submittals, plus routine on-site reviews by knowledgeable observers.

3.1.4 Proprietary Specifications

Proprietary specifications identify the desired products by manufacturer's name, brand name, model number, type designation, or other unique characteristics. When a manufacturer's name is not stated, a specification is considered proprietary when the product specified is available from only one source. Advantages of proprietary specifications are as follows:

- Product selection can be closely controlled.
- More detailed and complete drawings can be prepared based on precise information obtained from selected manufacturer's data.
- Reduced cost and time benefits may be obtained from use of shorter specifications and reduced drawing production effort.
- Bidding may be simplified by narrowing competition and removing product pricing as a major variable.

Disadvantages of proprietary specifications include the following:

- Competition for products is reduced or eliminated.
- Products may be specified with which the contractor may have had little or an unfavorable experience.
- Certain products and manufacturers may be favored over others.
- An error might occur when specifying model or product designations.

3.1.4.1 Closed and Open Proprietary Specifications

The fundamental distinction between closed and open proprietary specifications concerns substitutions. Closed proprietary specifications generally prohibit substitutions, whereas open specifications permit them. Different specifying objectives are served by each method. In the closed specification, the A/E is certain that the named products will be provided. In open specifications, the named product serves only to indicate the desired properties and acceptable level of requirements.

Closed Proprietary Specification

- Only one product is named.
- Several products may be named as options.
- There can be no substitutions.

Open Proprietary Specification

- Prices are requested for specified alternative products.
- Substitutions and cost adjustments may be proposed by the bidders.
- Products are allowed as substitutions after approval by A/E.

3.1.4.1.1 Closed Proprietary Specifications The closed proprietary specification permits the design to be completed to a high level of detail. This reduces variables and promotes accurate pricing; however, it does not offer protection against possible higher costs. The supplier of a specified proprietary product could take unfair advantage of being the sole source and increase the price of the product. Another difficulty occurs when a contractor is obliged to use a product with which it has had an unsatisfactory experience, such as slow delivery, lack of proper technical services, or difficult payment arrangements. Claims of collusion between the A/E and a sole-source supplier could be raised.

A closed proprietary specification can list one product or name several products as options. In either case, substitutions are not allowed. The control of the products and substitutions is accomplished by cross-reference in the instructions for procurement in Section 01 25 13—Product Substitution Procedures, which provides statements similar to the following:

- *One Product.* Where a product is specified by naming only one manufacturer, no substitute product will be considered. Bids must be based on the named product.
- *Optional Products.* Where products are specified by naming several manufacturers, no substitutes will be considered. Bids must be based on one of the named products.

Under the optional product approach, the successful contractor may be required to submit a product selection list at the time the bids are received or within a few days after the bid opening.

This method of listing several optional products overcomes many of the objections raised when only one product is specified. If at least three products are named, competition may be achieved; however, the A/E must ensure that the named products are equivalent and acceptable.

3.1.4.1.2 Open Proprietary Specifications Open proprietary specifications may alleviate the problem of overpriced sole-source items.

Requested Alternates. There are several ways that proprietary specifications can be opened to allow alternate products. One method is to request proposals for alternate products. This form of proprietary specification defines the materials in the same way as a closed specification (i.e., only one brand is named for each material or item of equipment specified). Alternates to the specified products are named in the specification. The proposals submitted must be for the base-bid items, but the bidder may list prices for specified alternates. Space for quoting alternates should be provided on the bid form. A statement similar to the following might be used in the instructions for procurement:

- Where a product is specified by naming only one manufacturer, substitute products will not be considered.
- Where alternates to the base-bid products are requested, bidders may list prices for alternates on the bid form. Bid price for each alternate product must include amount required to incorporate alternate product in the project. Requests for additional money for alternates after execution of agreement will not be considered.

This allows the bidder/proposer to select alternates and quote prices for preferred items. It may or may not be mandatory for the bidders/proposers to quote on all requested alternates.

The terms *alternate* and *substitute* are often misused. An alternate is something that is named for which alternative pricing is requested; a substitute is something that is requested to replace an item as specified.

Proposed Substitutions. This form of open proprietary specification is prepared in much the same way as a specification requesting alternates. However, no alternates are named in the specifications. The bid/proposal must be based on the specified materials, but the bidder/proposer is permitted to submit requests for substitutions, provided the bidder indicates the difference in cost that will result if the substitutions are accepted. Substitutions must include product name, nomenclature, and name of manufacturer, along with complete specifications and descriptive data. If the number of substitutions is unlimited, the task of evaluating and analyzing bids/proposals may become quite large.

A page should be provided with the bid form for listing proposed substitutions. The following paragraphs in the instructions to bidders help to eliminate the indiscriminate listing of products that the A/E must evaluate and help to ensure that the proposed substitutions will be worth considering:

- Bids shall be based on the specified products. However, bidders are encouraged to quote on substitute products by listing them on the substitution page of the bid form and by indicating the cost increase or decrease. Substitutions will not be considered after award of contract.
- The specified products have been used in preparing the drawings and specifications and thus establish minimum qualities that substitutions must meet to qualify as acceptable.
- Proof of equality rests with the bidder and adequate supporting information must accompany the bid.
- The bid price for each proposed substitution shall include all costs required to incorporate the substitution into the project. Later requests for additional money for substitutions will not be considered.

Controlled Substitutions. In this type of open proprietary specification, specific products are named but substitutions are allowed under procedures specified in Section 01 25 13—Product Substitution Procedures. A requirement may be met with the specified item or by a similar product that is not necessarily identical but that is alike with respect to performance. This method saves time during the development of a specification because only one product needs to be investigated and specified for each requirement.

The principal problem associated with this type of proprietary specification is that attempts are often made to substitute materials of different characteristics or requirements than those specified. This can occur when the time to submit proposed substitutions is improperly or inadequately specified. A weak specification is one that allows the contractor to propose substitutions after contract award. This practice leads to additional bid shopping and pressure on the A/E to accept many substitutions. The owner may sometimes end up with substitute products with different characteristics.

The control of the A/E can be enhanced if the following requirements are also included:

- *Substitution* requests from bidders are in writing and requests from manufacturers and suppliers are not considered.
- A *definite* deadline for the submittal of requests for substitutions is established.

- *Submittal* of supporting data by the bidders/proposers is required.
- A *written* acceptance of substitutions is issued to all bidders/proposers by an addendum.

Establishing the deadline for requests is the prerogative of the A/E. For many projects, the established date for indicating accepted substitutions will determine the time limit for product substitution requests. At least three days between deadline and addendum publication should be allowed for proper consideration.

The requirements and procedures for requesting substitutions are best specified in Section 01 25 13—Product Substitution Procedures. This section can be used to define limitations on substitutions, contractor's representation in requesting substitutions, method for requesting substitutions, and submittal procedures. While some government agencies require that proprietary names used in the specifications be followed by phrases such as "or approved equal," "or equivalent," and "or equal," this is redundant if the project manual contains a comprehensive Section 01 25 13. The Construction Specifications Institute (CSI) principle of "say it once" again applies.

The instructions to bidders should state times for submitting requests for substitutions during the bidding period; however, they should cross-reference to Section 01 25 13 for other requirements and procedures.

The following statement may be used in the instructions to bidders for this purpose:

> The specified products establish minimum requirements that substitutions must meet to be considered acceptable. To obtain acceptance of unspecified products, submit written requests at least 10 days before the bid date. Requests received after this time will not be considered. Refer to Section 01 25 13—Product Substitution Procedures for requirements and procedures for requesting substitutions.

The proprietary specification may be extended by including descriptive requirements of the specified products in addition to the brand name and nomenclature. Substitutions must meet these requirements. When the characteristics of products are specified in this manner, they establish a basis for acceptance or rejection of proposed substitutions within the specifications.

3.1.4.2 Selection of Proprietary Methods

There are many useful proprietary techniques that consider the bidder's/proposer's right to select products. Closed methods generally give the bidder/proposer little or no choice. Open methods give the bidders/proposers a wide choice.

The acceptance of substitutions opens the project to the possibility of accepting unknown and perhaps inferior products. If the specified item is part of a vital system, the situation may call for a proprietary specification prohibiting substitutions entirely. However, if keen competition and low construction cost are more important than completely assured performance, then substitutions should be permitted.

Open specifications generally place a greater workload on the A/E, increasing in direct proportion to the degree of choice available to the bidders/proposers. From the standpoint of reducing the design and specifying effort, the closed proprietary method is best. Because only one product is named for each application, the drawings and specifications can be completed quickly and precisely. There is no need to accommodate several combinations or construction options.

When substitutions are allowed, competition is keenest. Material costs are nearly certain to be less, but this method is conducive to permitting products of lesser requirements.

The apparent savings can easily be lost through subsequent high maintenance and replacement costs. These considerations reinforce the conviction that the A/E should carefully word the procurement documents to provide a definite, controllable method for exercising review and approval over what is accepted as a substitution.

A proprietary designation establishes a basis for determining quality, including performance, appearance, and cost. A proprietary name also establishes that the designer has considered the characteristics of the specified product, has decided upon its incorporation into the project, and has used its dimensions in the drawings. Any coordination effort resulting from the use of a contractor proposed substitution should be the responsibility of the contractor.

3.1.4.3 Nonrestrictive Specifications

The federal government and some other public authorities restrict the use of proprietary or exclusionary specifications except under special circumstances. A typical public law—in this case, the one providing for federal participation in the financing of wastewater treatment plants—includes the following language:

> Reference in the specifications to equipment, material, articles, or patented processes by trade name, make, or catalog number shall be regarded as establishing a standard of quality and shall not be construed as limiting competition. The Contractor may, at its option, use any equipment, material, article, or process that, in the judgment of the Contracting Officer, is equal to that named in the specifications, unless otherwise specifically provided in this contract.

Nonrestrictive specifications may be developed from descriptive, performance, reference standard, or proprietary specifications, or from a combination of these four methods. Descriptive and performance specifications can easily be used for nonrestrictive specifications. However, care must be taken to ensure that descriptive and performance requirements can be met by several manufacturers or suppliers. Failure to do so will make the specifications restrictive, even though a proprietary name has not been used.

A proprietary specification masquerading as a performance specification is not nonrestrictive. The A/E may delete the identifying name of the product and then list the salient qualities verbatim from the manufacturer's literature as product requirements. Without any ranges of characteristics and performance being given, this method results in a fully proprietary, closed specification and is not a suitable solution for a nonrestrictive project.

Several methods of converting a closed proprietary specification into a nonrestrictive specification have been used by A/Es practicing in public works fields. Public agencies usually require a minimum of three brand names followed by the words "or equal," along with descriptions of the salient characteristics of the item specified to provide a basis for judging equality.

Another method is the appendage of "or equal" to every manufacturer's name. This is awkward and redundant. A single paragraph in Division 01 is more effective than repetition of "or equal" throughout the specifications. A paragraph covering substitutions should be included in the instructions to bidders and in Section 01 25 00—Substitution Procedures. The following paragraph is typical:

> Whenever a product is specified by using a proprietary name or the name of a particular manufacturer or vendor, the specific item mentioned shall be understood as establishing type, function, dimension, appearance, and quality desired. Other manufacturers' products might be accepted, provided sufficient information is submitted to allow the A/E to determine that products proposed are equivalent to those named.

Chapter 4
Formats

4.1 Introduction to Formats

It is not practical to include sufficient notes on construction drawings to describe in complete detail product quality and installation methods. Separate written descriptions, referred to as *specifications,* more effectively communicate this type of information. The drawings and specifications are complementary.

Specifications define the qualitative requirements for products, materials, workmanship, and administrative requirements upon which the project is based. A simple definition of the word *specify* is to describe or define in detail. A specification is a precise statement describing the characteristics of the work. Whether specifying brick wall, curtain wall assembly, buildings, or bridges, the specifications must describe completely the desired results so that the work can be constructed in accordance with the contract requirements. Specifications in this context are different from other types of technical data and product descriptions available from project manufacturers that may also be referred to as specifications. Specifications are not just descriptions of products but detailed descriptions of the results of products with effort applied to these products to produce a work result—for example, the product *carpet* versus the work result of *carpeting.*

A number of organizational standards, referred to as *formats,* have been developed to organize all of the information that needs to be addressed from the inception of a project throughout its life cycle. These formats can each be employed for a variety of applications; the chief advantage they carry is a standardized means of organizing, storing, retrieving, and communicating this enormous body of data. Some more detailed descriptions of the characteristics and uses of these formats follow.

4.2 Hierarchy of Formats

The various formats establish an organizational hierarchy. Figure 4.1 illustrates these relationships. *OmniClass™* Construction Classification System (OCCS) is a multitable system for organizing information used by the architectural, engineering, and construction industry. The OCCS is the basis for deriving relational applications and will support and empower the transfer and use of information in the construction marketplace, ultimately serving all participants who work to sustain the built environment throughout the entire life cycle of a facility.

UniFormat™ organizes construction information based on the elements of a facility, otherwise known as systems and assemblies. These systems and assemblies are characterized by their function without identifying in detail the products that compose the system.

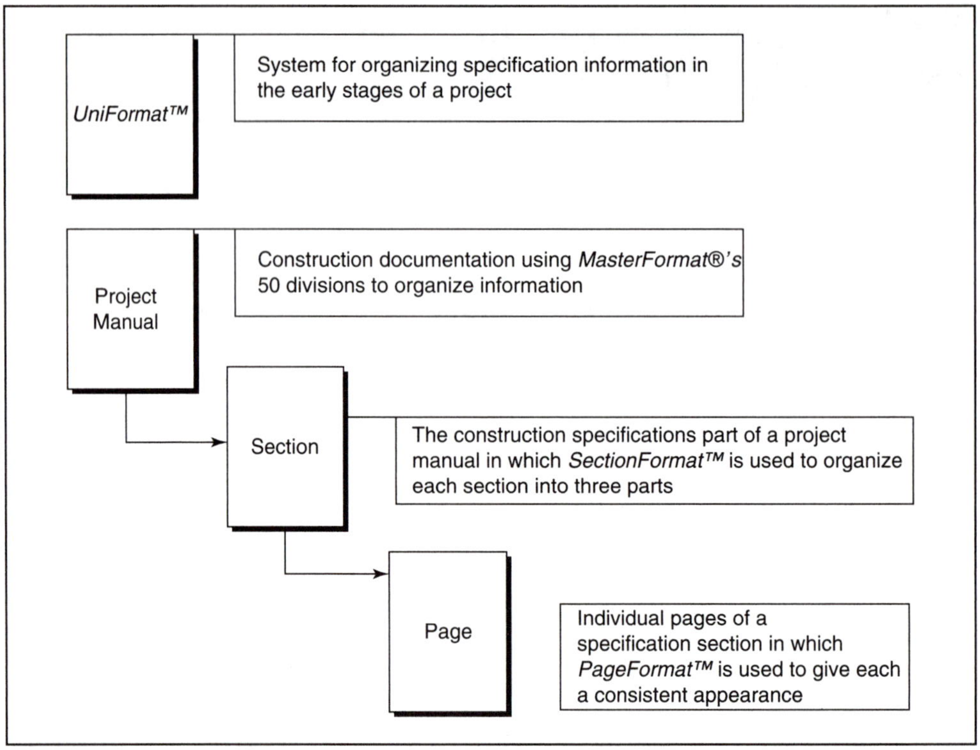

Figure 4.1
Hierarchy of the Formats

UniFormat™ is chiefly employed in the early design of a project, for preliminary project descriptions, and for performance specifying, and is frequently used for cost estimating.

MasterFormat® breaks the same information down by work results or construction practices that result from a combination of products and methods, and is utilized more often during the later design and construction stages of a project, but information organized by *MasterFormat®* is also used throughout the facility life cycle. The main purpose of *MasterFormat®* is to organize the project manual, reference keynotes, and detailed cost estimates.

Both *MasterFormat®* and *UniFormat™* are also designed to work within the OCCS, a multifaceted pure classification system designed to provide the capability for wide-ranging information storage and retrieval applications.

MasterFormat® establishes the organizational structure for the documents and sections within a project manual, each with its unique number and title. A section is further divided into the three PARTs defined in *SectionFormat™*. *PageFormat™* provides a consistent format for each page in a specification section. This simple hierarchy helps the architect/engineer (A/E) place information within the project manual and helps the user of the project manual find the information. For example, if a contractor or supplier wants to know what types of windows are required for a project, it is a simple process to find the answer.

- What is the general category of information? Division 08—Openings
- What is the specific category of information? Section 08 52 00—Wood Windows
- What is the detailed information? PART 2—PRODUCTS, Windows: horizontal sliding, plastic clad, wood units with double glazing

4.2.1 OmniClass™

OmniClass™ is a multi-table framework for organizing all forms of information used or encountered in the construction industry. The 15 *OmniClass™* classification tables can be used together or individually to manage information about all aspects of the facilities that comprise the built environment over their entire life cycle. *OmniClass™* concepts are derived from internationally accepted standards developed by the International Organization for Standardization (ISO) and supported by the International Construction Information Society and the International Alliance for Interoperability.

These concepts are described in detail in two ISO standards in particular:

- ISO 12006–2, Organization of Information about construction works—Part 2: Framework for classification of information, and ISO/PAS (Publicly Available Specification).
- ISO 12006–3, Organization of Information about construction works—Part 3: Framework for object-oriented information exchange, which, when taken together, define bases for organizing, managing, and transferring information about the built environment. These standards were also followed in the development of Uniclass: Unified Classification for the Construction Industry, a faceted classification system similar to *OmniClass™*, used in the United Kingdom in 1997. Additional ISO-compatible classification systems are being developed in other countries.

The scope of *OmniClass™* is designed to address objects of every scale through the entire built environment. *MasterFormat®* and *UniFormat™* address specific topics from specific viewpoints; the scope of *OmniClass™* is much broader. Where *OmniClass™* tables address similar topics, these legacy formats are incorporated within the scheme; *MasterFormat®* is incorporated as the basis of the *OmniClass™* table for work results, and *UniFormat™* is the basis of the table for elements.

4.2.1.1 Basis of Organization

The *OmniClass™* tables are organized by segregating types of information about the built environment into a set of discrete, coordinated tables. The information contained in each table exists and is organized based on a specific facet, or view, of the information for the subject matter addressed by that table.

OmniClass™ is composed of 15 tables, each representing a different broad facet of construction information. Each table can be used as a stand-alone table for the classification of a particular type of information or can be combined with terms from other tables to classify complex topics and do so with greater specificity than any table alone could provide.

The *OmniClass™* tables are:

Table 11—Construction Entities by Function
Table 12—Spaces by Function
Table 13—Construction Entities by Form
Table 14—Spaces by Form
Table 21—Elements
Table 22—Work Results
Table 23—Products
Table 31—Phases

Table 32—Services
Table 33—Disciplines
Table 34—Organizational Roles
Table 35—Process Aids
Table 41—Information
Table 42—Materials
Table 49—Properties

Entries in the *OmniClass™* tables are arranged in a hierarchical fashion. Tables are organized from broader high-level concepts to more specific lower-level, or granular, concepts. The *OmniClass™* tables taken as a whole also comprise an implied hierarchy of subjects, a progression from the general to the specific, with tables that deal conceptually with whole construction entities coming before the tables that dissect those objects into their composite elements, viewed from a variety of conceptual facets. Numbering of the entries in the *OmniClass™* tables consists of pairs of numbers for each level of hierarchy represented. The hierarchical organization of the *OmniClass™* tables allows for classification of topics at differing degrees of detail or extent. A system of signs, or delimiters, is used to combine the numbers that represent entries from different tables, allowing classification of more complex objects with more specificity.

4.2.1.2 Applications

OmniClass™ tables address topics and activities in construction information. The tables are structured as a faceted system that allows the tables to be used independently or in combination to organize hard-copy information libraries and projects as well as structure information in databases for storage and retrieval. At the simplest level, a single *OmniClass™* table can be used to organize a small collection of similar material; for instance, a small collection of product literature could be organized according to Table 23 — Products. For a larger collection, designators from more than one table can be used to qualify and refine entries, enabling information to be recalled according to user needs with far greater precision. For classifying large, complex collections of information, designators from multiple tables can be combined using *OmniClass™* delimiters to define conceptual relationships between table entries and classifying a wide variety of concepts contained within various data sources.

4.2.2 *UniFormat™*

UniFormat™ is a uniform classification system for organizing preliminary construction information into a standard order or sequence on the basis of systems and assemblies. By establishing a uniform list of identifiers and titles, *UniFormat™* promotes standardization and facilitates the retrieval of information. *UniFormat™* can be used to organize Preliminary Project Descriptions, preliminary cost estimates, and drawing detail filing in the early stages of a facility's life cycle. *UniFormat™*organizes construction information based on physical parts of a facility called systems and assemblies. These systems and assemblies are characterized by their function without identifying the products that compose them. Systems and assemblies present a view of a proposed facility separate from the view presented by a breakdown of building materials, products, and activities.

4.2.2.1 Basic Organization

UniFormat classifies information into nine Level 1 categories:

Project Description
Introduction
A Substructure
B Shell
C Interiors
D Services
E Equipment and Furnishings
F Special Construction and Demolition
G Building Sitework
Z General

These nine categories can be used to arrange preliminary project descriptions and preliminary cost information. Category Z, General, is designated by the last letter of the alphabet so the classification can expand beyond building construction. When the list is so expanded, this category will remain last. When included in project manuals, elemental (system and assemblies) performance specifications can be located in *MasterFormat®* under Section 01 80 00—Performance Requirements, in an order taken from these basic *UniFormat™* categories.

The *UniFormat™* numbering system is as follows:

A SUBSTRUCTURE
Level 1
A10 Foundations
Level 2
A1010 Standard Foundations
Level 3
A1010.XX XX XX Wall Foundations
Level 4
A1010.XX XX XX Continuous Footing
Level 5

Titles in Levels 1 through 3 can be applied to most preliminary project descriptions and preliminary cost estimates. Levels 4 and 5 are available for use on detailed, complex projects. Level 4 and 5 titles and detailed listings provide a checklist to ensure comprehensive and complete application of *UniFormat™*.

UniFormat™ uses a hierarchical organization of titles for systems and assemblies by employing two separate rules of classification. These two rules are specialization and decomposition.

At Level 1, the hierarchy comprises the major categories of construction information separated by their special function. These special functions include the nine Level 1 categories. The letters and titles of Level 1 categories are fixed and should not be changed in application.

The first category, Project Description, does not have a letter designation. This allows it to appear first in the list of titles and distinguishes its contents from the other titles. Project Description is not a building system or assembly. Instead, it is a collection of basic information that allows users to give readers an introduction to the project before reading details of systems and assemblies. Project Description contains bidding, proposal, and contract requirements and allows *UniFormat™* to be used as a stand-alone contracting structure for construction projects, especially design-build applications.

UniFormat™ divides Level 1 categories into classes of information by separating the categories into the constituent parts that compose them. This method of classifying is known to information specialists as *decomposition.*

Level 2 classes carry the letter of their parent category, plus a two-digit number. Alphanumeric designations and titles of Level 2 classes are also fixed and should not be changed in application.

Levels 3 and 4 are developed by further subdividing or decomposing Level 2 classes. These subclasses carry the alphanumeric designation of their parent category and class, plus another two-digit number.

Alphanumeric designations and titles of Level 3 subclasses are also fixed and should not be changed in application.

Alphanumeric designations for Level 4 are unassigned within a group of Level 3 subclasses, and number extensions are assigned by the user. The user-assigned number should carry the alphanumeric designation of the parent Level 3 subclass, plus a two-digit number, or the corresponding *MasterFormat®* number.

Titles presented below Level 4 are Level 5 subclasses, developed by presenting specialized design solutions of their parent Level 4 subclasses. Particular materials may be identified (e.g., concrete and steel) to differentiate one specialized solution from another. Titles at Level 5 are examples of information included in their Level 4 subclass.

No alphanumeric designation is assigned. Extensions may be assigned by the user. The user-assigned number should carry the alphanumeric designation of the parent Level 4 subclass, plus a two digit number, or the corresponding *MasterFormat®* number. Level 5 and, sometimes, Level 4 titles begin to correspond to the concepts contained in *MasterFormat®* numbers and titles. The numbering system can expand to allow for additional numbers and titles at any level. This allows *UniFormat™* to expand in the future and allows users to add titles and numbers for subjects not in the current edition. User-defined numbers and titles that fit within the established framework of *UniFormat™* can be added. Figure 4.2 is a list of Level 1 and 2 alphanumeric designations and titles from *UniFormat™*.

4.2.2.2 Applications

Preliminary Project Description *UniFormat™* provides a system to describe a project by its basic systems and assemblies before the particular materials and methods have been chosen. A/Es use this document to organize project descriptions and help estimators formulate early cost estimates.

This format is used by many federal agencies for their cost classification system.

Refer to the discussion on Preliminary Project Descriptions and Outline Specifications in Project Delivery Practice Guide for more information regarding the use of *UniFormat™* in organizing preliminary project descriptions.

The basic underlying ideas behind the concept of the Preliminary Project Description are:

- Written descriptions of the Schematic Design should be organized around systems and assemblies that correlate to prevailing industry methods of cost estimating for this phase.

Introduction
- 10 Project Description
- 20 Owner Development
- 30 Procurement Requirements
- 40 Contracting Requirements

A Substructure
- A10 Foundations
- A20 Subgrade Enclosures
- A40 Slabs-On-Grade
- A60 Water And Gas Mitigation
- A90 Substructure Related Activities

B Shell
- B10 Superstructure
- B20 Exterior Vertical Enclosures
- B30 Exterior Horizontal Enclosures

C Interiors
- C10 Interior Construction
- C20 Interior Finishes

D Services
- D10 Conveying
- D20 Plumbing
- D30 Heating, Ventilation, And Air Conditioning (Hvac)
- D40 Fire Protection
- D50 Electrical
- D60 Communications
- D70 Electronic Safety And Security
- D80 Integrated Automation

E Equipment And Furnishings
- E10 Equipment
- E20 Furnishings

F Special Construction And Demolition
- F10 Special Construction
- F20 Facility Remediation
- F30 Demolition

G Sitework
- G10 Site Preparation
- G20 Site Improvements
- G30 Liquid And Gas Site Utilities
- G40 Electrical Site Improvements
- G50 Site Communications
- G90 Miscellaneous Site Construction

Z General
- Z10 General Requirements
- Z70 Taxes, Permits, Insurance And Bonds
- Z90 Fees And Contingencies

Figure 4.2
UniFormat Levels 1–2

- Written descriptions should allow design professionals to provide sufficient information for cost estimating without the necessity of making final design decisions.
- Written descriptions should document qualitative requirements for the project appropriate to the level of decision-making and detail in the design.
- Using an industry-standard organizational format provides a checklist to help design teams make sure all appropriate subjects are included.

In addition to providing more detail for preparing Preliminary Project Descriptions, *PPDFormat* has additional material to respond to the following changes in the industry since publication of The Project Resource Manual and the library of Practice Guides:

- Coordination with the 2010 edition of *UniFormat™*
- Building Information Modeling (BIM)
- Integrated Project Delivery (IPD)

Preliminary Cost Estimates *UniFormat™* also serves as the basis for the organization of preliminary cost estimates and parallels the organization of preliminary project descriptions. Familiarity with *UniFormat™* allows users to easily relate a project description to cost data. There are commercially available cost data systems based on *UniFormat™*.

Drawing Detail Filing *UniFormat™* also serves as a system for filing and retrieving drawing details. Because most details include multiple materials as classified by *MasterFormat®*, it is difficult to file details under that system. *UniFormat™*, by contrast, is organized by elements, systems, and assemblies that relate to the element, system, or assembly depicted in a detail. *UniFormat™* is used for the internal filing and retrieval of details in the office producing or using the details; it is not proposed for use in the identification of details on a set of drawings. Other potential uses of *UniFormat ™* include scheduling, value analysis, building performance, recording of design data, structuring of codes, and monitoring and management of design programs and costs.

4.2.3 *MasterFormat®*

MasterFormat®, jointly developed by the Construction Specifications Institute (CSI) and Construction Specifications Canada, is an organizational structure providing numbers and titles for the variety of subject matter necessary for the construction, operation, and maintenance of a facility. *MasterFormat®* provides a system of six-digit and eight-digit numbers and titles for organizing construction information into a standard order or sequence. By establishing a master list of numbers and titles, *MasterFormat®* promotes standardization, facilitates the placement and retrieval of information, and improves construction communication.

4.2.3.1 Basic Organization

The numbers and titles in *MasterFormat®* are organized into two groups:

- Procurement and Contracting Requirements
- Specifications

The first group, Procurement and Contracting Requirements, is the location for introductory information and procurement information as well as the location for information defining the relationships, responsibilities, and processes for construction. The second group, Specifications, provides the locations to describe administrative requirements and the physical aspects of construction.

4.2.3.2 Divisions

MasterFormat® arranges related construction practices, or work results, into a series of Level 1 titles called *Divisions.* Several of these divisions are without contents or titles; these are numbers reserved for potential future expansions in content. Central to the arrangement and use of contents of the divisions in the groupings in *MasterFormat®*, and the sections that make them up, is the notion that all of the different types of construction are addressed equally. The more basic, or common, divisions are generally placed nearer to the beginning of *MasterFormat®*. These contain work results likely to be specified in all types of construction.

For example, all new building construction projects have contractual requirements and general requirements, and are constructed using some kind of material. Later in *MasterFormat®* come the sub-groupings of divisions that contain work result sections applicable only to specific types of projects (building construction, heavy civil work, process plant construction, and the like).

The only exception to this general rule is for civil work; all projects reference sections in the "Site and Infrastructure" subgroup, located in the "30-series" of divisions, for their civil work.

The basic organizational structure of the *MasterFormat®* groups and subgroups, including division numbers and titles, is as follows:

Procurement and Contracting Requirements Group
- Division 00—Procurement and Contracting Requirements

Specifications Group

General Requirements *Subgroup*:
- Division 01—General Requirements

Facility Construction *Subgroup*:
- Division 02—Existing Conditions
- Division 03—Concrete
- Division 04—Masonry
- Division 05—Metals
- Division 06—Wood, Plastics, and Composites
- Division 07—Thermal and Moisture Protection
- Division 08—Openings
- Division 09—Finishes
- Division 10—Specialties
- Division 11—Equipment
- Division 12—Furnishings
- Division 13—Special Construction
- Division 14—Conveying Equipment
- Division 15—(Reserved for future expansion)
- Division 16—(Reserved)
- Division 17—(Reserved)
- Division 18—(Reserved)
- Division 19—(Reserved)

Facility Services *Subgroup*:
- Division 20—(Reserved)
- Division 21—Fire Suppression
- Division 22—Plumbing
- Division 23—Heating, Ventilating, and Air-Conditioning
- Division 24—(Reserved)
- Division 25—Integrated Automation

Division 26—Electrical
Division 27—Communications
Division 28—Electronic Safety and Security
Division 29—(Reserved)

Site and Infrastructure *Subgroup*:
Division 30—(Reserved)
Division 31—Earthwork
Division 32—Exterior Improvements
Division 33—Utilities
Division 34—Transportation
Division 35—Waterway and Marine Construction
Division 36—(Reserved)
Division 37—(Reserved)
Division 38—(Reserved)
Division 39—(Reserved)

Process Equipment *Subgroup*:
Division 40—Process Integration
Division 41—Material Processing and Handling Equipment
Division 42—Process Heating, Cooling, and Drying Equipment
Division 43—Process Gas and Liquid Handling, Purification and Storage Equipment
Division 44—Pollution and Waste Control Equipment (revised for MF2010)
Division 45—Industry-Specific Manufacturing Equipment
Division 46—Water and Wastewater Equipment (revised for MF2010)
Division 47—(Reserved)
Division 48—Electrical Power Generation
Division 49—(Reserved)

4.2.3.3 Numbering

MasterFormat® uses three pairs of digits as a numbering structure and allows for an additional optional pair of digits for expansion.

Following is an example of how the numbering system is organized.

Section 03 20 00—Concrete Reinforcing

The first two digits represent Level 1 Divisions, in this case Division 03—Concrete. The next pair of numbers, in this case 20, represents Level 2, and the third pair, 00, represents Level 3 numbers. As each level of classification is represented by pairs of numbers, there is room to provide flexibility and room for expansion if required. Most Level 4 numbers are not published within *MasterFormat*®, and individual users may wish to use the Level 4 numbers for specific requirements.

Documents in the Procurement and Contracting Requirements group, Division 00, are not specifications; however, *MasterFormat*® assigns standard locations and numbers for these documents because they are usually included in the project manual. The remainder of the project manual, the titles in the Specifications Group, is made up of specification sections. Divisions 01 through 49 are specifications and form the Specifications Group. Divisions 02 through 49 are Technical Specifications.

It is not the intent of *MasterFormat*®to assign a single standard location for materials, products, or assemblies. Rather, *MasterFormat*®is a list of titles that represent construction practices or "work results" that result from the application of skills and procedures to the materials, products, or assemblies. Thus, it is possible that different applications of a product would be found in different locations. Plywood panels, for example, can be placed in Division 03 for their function as concrete forms, in Division 06 for their application as interior finish paneling, in Division 07 for their function as exterior siding for weather protection, or in Division 27 as a telephone board. Coupled with a specifier's construction experience and knowledge, *MasterFormat*®'s keyword index makes organizing a project specification logical and effective. *MasterFormat*®provides a fixed yet expandable framework for organizing specifications, provides a standard sequence for arranging specification sections, and enables easy retrieval of information.

4.2.3.4 Using *MasterFormat*®

A random designation system does not fulfill the needs of those who wish to standardize portions of their specifications, nor does it meet the requirements of those who use electronic media for data retrieval and automated processing. *MasterFormat*®provides an order that helps everyone involved in a construction project avoid this outcome and these potential communications disconnects.

4.2.3.5 Keyword Index

An alphabetical subject index is included in *MasterFormat*®as a convenient means of guiding the user in locating the proper grouping of titles for construction products and activities. The reference within the index is by the six-digit classification number that designates the Level 1, 2, or 3 titles.

4.2.3.6 Applications

MasterFormat®is an organizing system with the flexibility to meet the needs of any particular company, professional discipline, or type of project. It is used by A/Es to organize project manuals and to help users of project manuals find information in standard locations. It is used by estimators for organizing cost data, and it is suitable for either manual or electronic data management operations. It is used by A/E firms specializing in particular types of projects. *MasterFormat*®is also used by manufacturers and suppliers to organize and identify product literature and to market their products to users.

4.2.3.7 Specifications

MasterFormat®is most widely used for the purpose that drove its initial development: organizing specifications. Titles are provided in a logical sequence for the most common specification sections required for a construction project.

Use of the numbers and titles shown in *MasterFormat*® will allow easy cross-referencing within the project manual because a section will always have the same number and title. Sections can be written and reproduced at any time without renumbering the other sections in the project manual. This is one of the major advantages in the use of *MasterFormat*®.

Assembly of the final document in numerical sequence ensures the correct grouping of sections. For *contractors* and suppliers who are familiar with *MasterFormat*®, the retrieval of information during bidding and construction will be faster, easier, and more thorough than with a nonstandardized set of specifications.

The A/E using *MasterFormat*® can specify materials or systems work results, or a combination of both. A system may be specified in a single Level 2 section, or the general *system* may be specified in a Level 2 section and its components in several Level 3 sections. *MasterFormat*® has been developed to provide the A/E with a standard yet flexible system for organizing specifications and construction information.

MasterFormat® sections are not intended to correspond to the work assigned by the contractor to a *single* trade or subcontractor.

4.2.3.8 Data Filing

The system of numbers and titles serves for filing and retrieving technical data and product literature organized by work results. Data filing under *MasterFormat*® is *primarily* concerned with the rapid retrieval of technical data related to products or work results being specified. Because the same system is used for data filing as for organizing specifications, it is easy to relate the filed material to the specifications being written for a project.

MasterFormat® is used by many construction product manufacturers for their product literature, allowing their products to be related to the work result sections in which they are used.

4.2.3.9 Cost Classification

The number and title system serves as the basis for the organization of construction costs and parallels the organization of specifications. Familiarity with *MasterFormat*® and *UniFormat*™ allows users to relate a specification section to both product information and cost data.

4.2.3.10 Product Data

MasterFormat® has contributed significantly to standardization within the construction industry and is widely used to organize specifications, product literature, and cost information under a single classification system based on work results or construction practices. For a more detailed discussion on the use of *MasterFormat*®, refer to the *Introduction and Application Guide to MasterFormat®—Master List of Numbers and Titles for the Construction Industry.*

4.2.4 *SectionFormat*™

4.2.4.1 Basic Organization

SectionFormat™ provides a uniform approach to organizing specification text contained in a project manual. *SectionFormat*™ assists in the organization of specification sections by establishing a structure consisting of three primary parts. These parts organize text consistently within each section. Although the format was developed to assist with preparation of sections in Divisions 02 through 49, the principles also apply to sections in Division 01—General Requirements (refer to Figure 4.3). A section is intended to cover one portion of the project requirements. It describes particular materials, products, systems, or assemblies and their installation and particular administrative or procedural requirements. Individual sections with related items are grouped together under the appropriate divisions within *MasterFormat*®. Specific sections are included in a project manual specification only as needed to meet the project requirements. Specifications address the contractor. A section does not necessarily relate to the work accomplished by a single subcontractor. It is not the intent of the specifications to define the work of individual trades. Each contractor will divide the work differently among subcontractors. The contractor executes an agreement with the owner to construct the project. Responsibility for construction remains with the contractor regardless of how the work is divided among subcontractors and suppliers.

SectionFormat Outline

Standard Article titles in a section. **BOLD UPPERCASE:** Primary titles. Title Case: Subordinate titles that may be elevated to primary Article titles.

PART 1— GENERAL

SUMMARY
- Section includes
- Products Furnished [OR] Supplied but Not Installed under This Section
- Products Installed but Not Purnished [OR] Supplied under This Section
- Related Requirements

PRICE AND PAYMENT PROCEDURES
- Allowances
- Unit Prices
- Alternates [OR] Alternatives
- Measurement and Payment

REFERENCES
- Abbreviations and Acronyms
- Definitions
- Reference Standards

ADMINISTRATIVE REQUIREMENTS
- Coordination
- Preinstallation Meetings
- Sequencing
- Scheduling

SUBMITTALS
ACTION SUBMITTALS/INFORMATIONAL SUBMITTALS
- Product Data
- Shop Drawings
- Samples
- Certificates
- Delegated Design Submittals
- Test and Evaluation Reports
- Manufacturer Instructions
- Source Quality Control Submittals
- Field [OR] Site Quality Control Submittals
- Manufacturer Reports
- Sustainable Design Submittals
- Special Procedure Submittals
- Qualification Statements

CLOSEOUT SUBMITTALS
- Maintenance Contracts
- Operation and Maintenance Data
- Bonds
- Warranty Documentation
- Record Documentation
- Sustainable Design Closeout Documentation
- Software

MAINTENANCE MATERIAL SUBMITTALS
- Spare Parts
- Extra Stock Materials
- Tools

QUALITY ASSURANCE
- Regulatory Agency Sustainability Approvals
- Qualifications
  - Manufacturers
  - Suppliers
  - Fabricators
  - Installers/Applicators/Erectors
  - Testing Agencies
  - Licensed Professionals
- Certifications
- Sustainability Standards Certifications
- Preconstruction Testing
- Field [OR] Site Samples
- Mock-ups

DELIVERY, STORAGE, AND HANDLING
- Delivery and Acceptance Requirements
- Storage and Handling Requirements
- Packaging Waste Management

FIELD [OR] SITE CONDITIONS
- Ambient Conditions
- Existing Conditions

WARRANTY [OR] BOND
- Manufacturer Warranty
- Special Warranty
- Extended Correction Period

(*Continued*)

PART 2—PRODUCTS

OWNER-FURNISHED [OR] OWNER-SUPPLIED PRODUCTS
- New Products
- Existing Products

[SYSTEMS]/[ASSEMBLIES]/[MANUFACTURED UNITS]/ [EQUIPMENT]/[COMPONENTS]/[PRODUCT TYPES]/[MATERIALS]/[USER-DEFINED HEADING]

Manufacturers
- Manufacturer List
- Substitution Limitations
- Product Options

Description
- Regulatory Requirements
- Sustainability Characteristics

Performance/Design Criteria
- Capacities

Operation
- Operators
- Controls
- Operation Sequences

Materials

Assembly [OR] Fabrication
- Factory Assembly
- Shop Fabrication
- Assembly [OR] Fabrication Tolerances

Mixes

Finishes
- Primer Materials
- Finish Materials
- Shop Finishing Methods

ACCESSORIES

SOURCE QUALITY CONTROL
- Tests and Inspections
- Nonconforming Work
- Manufacturer Services
- Coordination of Other Tests and Inspections

PART 3—EXECUTION

INSTALLERS
- Installer List
- Substitution Limitations

EXAMINATION
- Verification of Conditions
- Preinstallation Testing
- Evaluation and Assessment

PREPARATION
- Protection of In-Place Conditions
- Surface Preparation
- Demolition/Removal

ERECTION/INSTALLATION/APPLICATION/ [USER-DEFINED PROCESS]
- Special Techniques
- Interface with Other Work
- Systems Integration
- Tolerances

[REPAIR]/[RESTORATION] REINSTALLATION

FIELD [OR] SITE QUALITY CONTROL
- Field [OR] Site Tests and Inspections
- Nonconforming Work
- Manufacturer Services

SYSTEM STARTUP

ADJUSTING

CLEANING
- Waste Management

CLOSEOUT ACTIVITIES
- Demonstration
- Training

PROTECTION

MAINTENANCE

ATTACHMENTS

END OF SECTION

Schedules
Tables
Illustrations
Forms

Figure 4.3
SectionFormat™ Outline

4.2.4.2 Parts

Written material within sections can be divided into three parts:

- *PART 1—GENERAL.* Describes administrative, procedural, and temporary requirements unique to the section. PART 1 is an extension of subjects covered in Division 01 and amplifies information unique to the section.

- *PART 2—PRODUCTS.* Describes products, materials, equipment, fabrications, mixes, systems, and assemblies that are required for incorporation into the project. Materials and products are included with their quality requirements.
- *PART 3—EXECUTION.* Describes installation or application, including preparatory actions and post-installation cleaning and protection. Site-built assemblies and site-manufactured products and system are included.

Sections are not intended to stand alone. They function with other portions of the procurement and contract documents and must relate to them.

For example, PART 1—GENERAL describes administrative and procedural requirements specific to the subject being covered. The requirements described in PART 1 should not duplicate statements that are contained in sections of Division 01. Each article and paragraph should supplement and be coordinated with the applicable sections of Division 01 to avoid repetition or conflicting requirements.

4.2.5 PageFormat™

There are advantages to standardizing the way information is presented on a page. A standard page format provides an orderly and uniform arrangement of text for each page of a specification section. The standard page format has three objectives:

- To present text clearly and at a density best suited for easy reading and rapid reference
- To provide an acceptable standard suitable for use in specifications throughout the construction industry
- To provide a flexible format compatible with most current production techniques and electronic software

Uniformity of presentation eases the tasks of preparation, review, and publication and saves the specification user time and effort.

Chapter 5 Agreements

5.1 Construction Agreements

Although often referred to as the *contract,* the construction agreement is only one of the various documents that make up the contract documents. The agreement is the document that legally obligates the signing parties. It binds the documents together firmly and permanently, while allowing for the addition of modifications, as illustrated in Figure 5.1. The agreement defines relationships and obligations between the signers and incorporates, by reference, all of the other documents that make up the contract documents: contracting forms; project forms; conditions of the contract (general and supplementary conditions); specifications; contract drawings; and revisions (addenda), clarifications, and modifications. The clauses and provisions of the general conditions for a construction contract are generally fixed and usually require only supplements to be adapted to a specific project. The agreement contains contractual elements that are subject to variation in each contract. These elements are:

- Date
- Identification of parties (the owner and the contractor or design-builder)
- Identification of the third parties (architect/engineer [A/E] and construction manager)
- Statement of work to be performed (extent or scope)
- Statement of the consideration or contract amount (cost, sum, or price)
- Statement of commencement and completion (time of performance)
- Signatures

Whether or not a standard form is used, the agreement has important legal consequences. No agreement form should be incorporated in construction documents without the owner's approval. In addition, standard agreement forms should not be prepared, completed, or modified without consulting the owner's legal counsel.

The provisions within standard agreement forms and respective general conditions are carefully interrelated. A modification in the language may create ambiguity or result in unintended legal obligations of the parties involved.

5.2 Project Delivery and Basis of Payment

Usually, an owner initiates a single construction contract or multiple-prime contracts. A construction agreement might be executed with a contractor, a construction manager as a contractor, or a design-builder. Each agreement contains specific text addressing the roles and responsibilities for the respective project delivery.

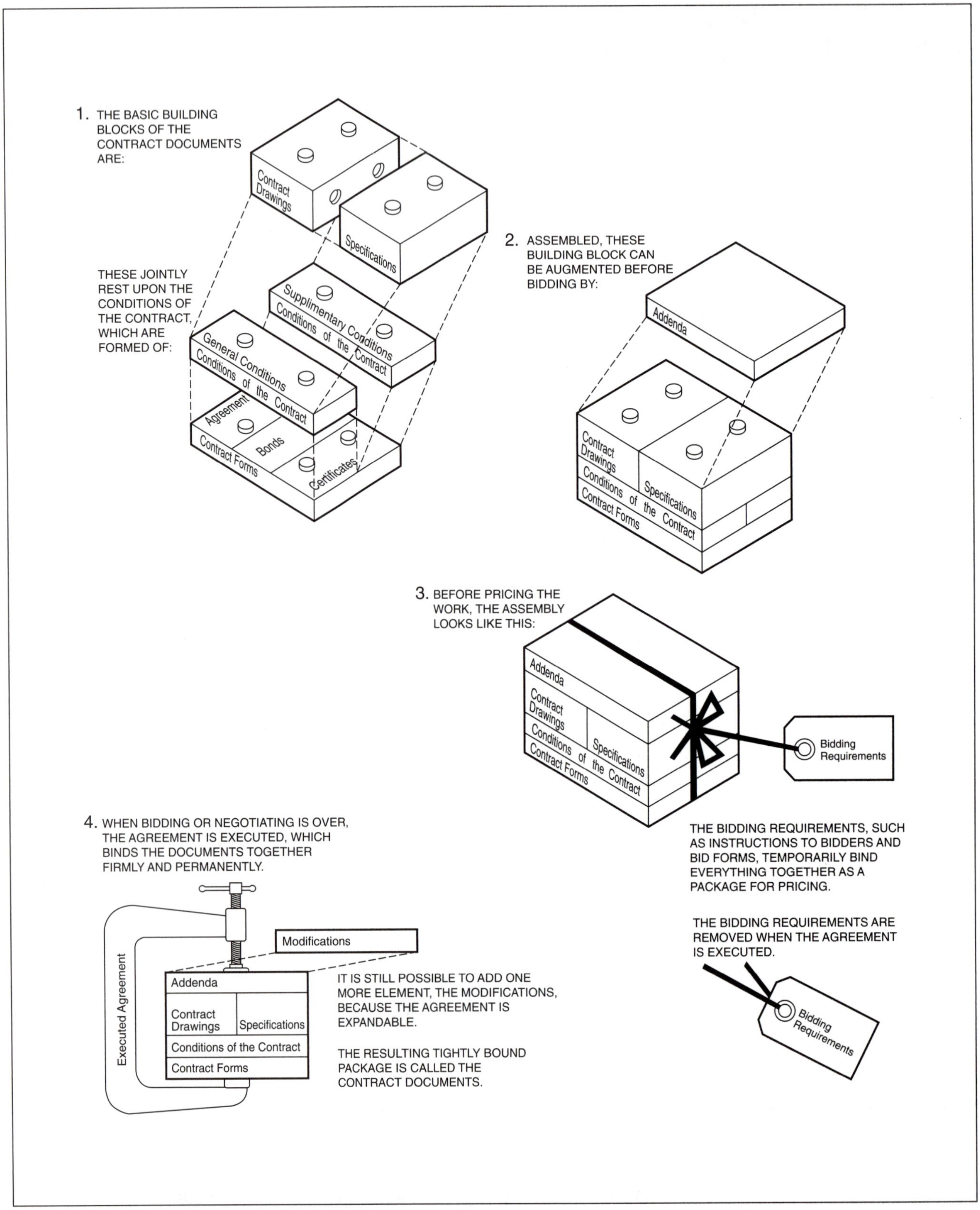

Figure 5.1

Graphic Representation of Relationship of the Documents Contained in the Agreement

In addition to the various types of project delivery, there are also different types of agreements with text for the basis of payment. The basis of payment for an agreement might include one of the following:

- Stipulated price or lump sum
- Unit price
- Cost-plus fee
- Cost-plus fee with guaranteed maximum price

Depending on the project delivery, the number of contracts, and the basis of payment, a specific form of agreement will be required that will state the relationships and obligations of the signing parties for the variations involved.

5.3 Standard Agreement Format

The following is the general organization of most standard agreement forms:

- *Preamble/Cover Page.* Stating the date, project, parties, and participants.
- *Articles.* Covenants that provide the basic essentials for a valid contract and define the intended rights and responsibilities of the signing parties.
- *Scope.* Extent of work to be performed.
- *Time.* Dates of commencement and completion or duration.
- *Cost.* Compensation or payment.
- *Affirmation.* A concluding statement, which precedes the signatures.
- *Signatures.* These are the signatures of the contracting parties. Also included are the signatures of witnesses and corporate seals, if required.

5.4 Standard Agreement Forms

On the basis of the project delivery method and the payment, the appropriate agreement form can be selected. Professional associations and government agencies have developed various standard agreement forms that may be modified for specific project requirements. This standardization of forms is desirable because it may save preparation time and offer reasonable assurance of a complete and accurate document. Through repeated use, forms with standardized clauses have become well understood by owners, A/Es, engineers, and contractors and are less subject to misinterpretation. However, their incorrect use may weaken a contract and in some instances might result in disputes.

The agreement form and the respective conditions of the contract should be reviewed carefully by the owner's legal counsel before modification to satisfy specific project requirements. The standard agreement forms and conditions of the contract contain wording that has been interpreted and adjudicated, and any modifications should be made with caution. Modifications to one document may require coordination with and modification to another. For example, modifications without proper coordination with the associated conditions of the contract may create conflicts and problems within the contract documents.

Various standard agreement forms have been developed and are in common use on construction projects. Some corporations and government agencies have prepared specific forms that must be used in their contracts.

Regardless of their origin, standard agreement forms contain common clauses that indicate contractual stipulations, including the following:

Cover Page/Preamble Introduces and includes, as a minimum, the date of the agreement and the identification of the project, and the names of parties and participants.

Identification of Parties Provides space for the legal names of the parties entering into the agreement, specifically the owner and contractor. Addresses and phone numbers would also be included to provide a location where proper notice (as required by various terms and actions) can be delivered.

Identification of Other Participants Identifies other participants who affect the agreement between the parties and may include the A/E for the project or a construction manager as adviser (CMa), and refers to the contract documents for the rights and authority assigned to them.

Contract Documents Defines and enumerates the contract documents and incorporates all of them by reference in the contract. It also provides for an accurate detailed enumeration of all written and graphic documents that comprise the contract. The list may make reference to documents (revisions, clarifications, and modifications) that might be issued after the effective date of the agreement.

Work of This Contract Describes the scope of the work, under the agreement, to be performed in accordance with the contract documents, including labor, materials, equipment, and services to fulfill the obligations. The work may be the complete project or a specific portion of the project when multiple prime contracts are indicated by the project delivery.

Time and Completion Establishes important time requirements:

- *Date of Commencement of the Work.* When time of performance is to be strictly enforced, the statement of starting time should be carefully considered to allow for review of all required preconstruction submittals. If a notice to proceed is designated in the agreement form, such notice would be issued after all matters relating to the contract and the construction operations have been determined.
- *Date of Substantial Completion of the Work.* Establishes a number of days (preferably calendar days) or a specific date. Substantial completion is the point in the progress of the work when the owner can occupy the project or a designated portion thereof for its intended use. The time requirements will have been met when the work is substantially complete, even if a few minor items may remain to be completed or corrected. There are occasions when delay, beyond the time limit stated in the contract documents, will cause damage to the owner. In such cases, the owner will be damaged by delay.
- *Liquidated Damages.* Establishes a fixed monetary amount per day, as a result of failure by the contractor to complete the work by a stipulated time or by a stated calendar date. The amount per day should be stated in the agreement and referenced in the procurement instructions and in the conditions of the contract. The owner, under advice from legal counsel, should determine both the type of damages to be covered or excluded and the amount per day that is stipulated due from the contractor. The inclusion of a liquidated damages clause in a contract is generally considered to be in lieu of actual damages for delay. If there is no liquidated damages clause, however, the contractor may still be liable for actual damages.

- *Incentives, Bonuses, and Penalties.* Establishes, if applicable, incentive, bonus, and penalty amounts if the project is completed early. Although this is not a required or a necessary condition when imposing liquidated damages, it is highly recommended if the owner will benefit from early completion.

Contract Sum/Price Establishes the contract sum (accepted bid or negotiated amount), including accepted alternates, and, when permitted, adjustments resulting from minor changes negotiated after receiving bids or prices. It is the amount to be paid for the performance of the contract.

- *Adjustment of Contract Sum.* Provides that the owner has the right to make changes to the work without invalidating the contract. Without such a provision, changes might not be permitted. Such changes might require appropriate adjustments in the contract sum.
- *Unit Prices.* Includes unit price items, if applicable, based on estimated quantities of work included from the bid form, or may include unit price items that are negotiated at the time of executing the agreement. Unit prices may be used to determine the value of changes in the work.

Payment Procedures Establishes requirements for preparing and processing applications for payment. It establishes time periods for submitting applications and receiving payment. This article might also establish an interest rate for monies not paid when due.

- *Due Dates for Payments.* Establishes dates mutually acceptable to the parties considering the time required to prepare an application for payment, for the A/E to check and certify payment, and for the owner to make payment within the time limits identified in the contract documents.
- *Work Completed.* Establishes provision that the contractor may be paid for work completed and in place, as well as for materials or equipment not incorporated in the work but delivered and suitably stored and protected on-site.
- *Materials and Equipment Stored Off-Site.* Establishes conditions if payment is permitted for materials and equipment stored off-site at a mutually agreed location, properly protected and insured, with clear title to owner.
- *Cutoff Date.* Establishes the date on which the work is evaluated for payment; it should normally be no fewer than 10 days prior to the payment date to allow for the A/E to assess the evaluation and issue a certificate for payment, and the owner to make payment within the time provided in the contract documents. The contractor may prefer an additional few days to allow time to prepare the application.
- *Payment at Substantial Completion.* Establishes the percentage of the contract sum that will be due upon substantial completion. The amount to be stipulated for the percentage retained at substantial completion should be decided by the owner and the owner's legal counsel with advice from the A/E.
- *Due Date for Final Payment.* Frequently established as 30 to 60 days after substantial completion, which would in most cases allow adequate time for final completion. However, it may be stipulated that this time period is subject to specific requirements of the project and, further, that this date is binding only if the work is in fact fully completed.
- *Progress Payments/Retainage.* Establishes a monthly date, mutually acceptable to the parties, for progress payments to be made. Additional paragraphs might be added to cover retainage percentages prior to and upon substantial completion.

Traditionally owners pay 90 to 95 percent of earned sums, retaining the remainder to ensure performance of the contract. The amount retained is often governed by local custom and, for public work, by state or federal regulations or statutory requirements. This procedure provides the owner with some degree of assurance that the contractor will perform the work expeditiously, but it also may increase project costs because a contractor will compensate for the amount of retainage in its price. A fair and effective practice is for the owner to place retained funds into interest-bearing escrow accounts. The funds remain under control of the owner until project completion, at which time the retainage, plus the interest accrued, is paid to the contractor.

Other customary procedures are to reduce the amount of retainage from 10 percent to 5 percent after the project is 50 percent complete, or retain 10 percent of payments due until the project is 50 percent complete and not retain on payments due thereafter. AIA documents include retainage provisions in the agreement and in the guide to the supplementary conditions. The Engineers Joint Contract Documents Committee (EJCDC) locates retainage provisions in the agreement only.

Miscellaneous Provisions States that where reference is made in the agreement to a provision of the general conditions or another contract document, the reference refers to that provision as amended or supplemented by other provisions of the contract documents. Miscellaneous provisions might also state that interest can be applied to payments under the contract documents that are due and unpaid, at a predetermined rate.

Termination or Suspension Establishes that the contract may be terminated or suspended by the owner.

Signatures Provides spaces for signatures of the parties to the agreement. The agreement form should be executed in sufficient quantities so that the owner and each party and participant have access to original agreement documents.

5.5 Inclusion of Agreement in Project Manual

Even though standard agreement forms, published by professional societies, other organizations, and agencies, will be utilized, the A/E should avoid including the agreement by reference only. It is more appropriate to include the actual agreement forms in the project manual. The possibility of referencing an incorrect or out-of-date form is also eliminated if the forms are bound into the project manual.

Chapter 6 Conditions

6.1 Conditions of the Contract

Conditions of the contract define basic rights, responsibilities, and relationships of the parties involved in the performance of the contract. These conditions are an inherent part of the owner-contractor agreement and are considered to be the *General Conditions* of the agreement. Limited services or abbreviated agreements normally contain the general conditions in the actual agreement. However, the involvement of other participants (e.g., architect/engineer [designer] or construction manager) and the delegation of responsibilities to subcontractors lead to the relatively public nature of the conditions of the contract. The need for review by those not signatory to the agreement has led to the use of conditions of the contract separate from the agreement. Conditions of the contract consist of the following:

- *General Conditions.* Usually in the form of published standard documents that include administrative provisions common to most construction contracts.
- *Supplementary Conditions.* Specially prepared to modify and expand the general conditions as needed to accommodate the unique requirements of a specific project.

The American Institute of Architects (AIA) and Engineers Joint Contract Documents Committee (EJCDC) electronic versions allow for direct modification of standard general conditions. The resulting project-specific general conditions contain both the original text and the supplementary modifications. In this case, there might not be separate supplementary conditions. Though not a signer of the owner-contractor agreement, the architect/engineer (A/E) commonly has duties and responsibilities described in the conditions of the contract and may assist in their preparation. A/E duties and responsibilities under conditions of the contract generally include construction administration. These activities are discussed in the *CSI Construction Contract Administration Practice Guide* and include:

- Acting as an owner's representative for matters concerning the work
- Visiting the construction site to become familiar with the work, guarding against defects and deficiencies, and evaluating progress
- Validating the progress of the work for the purpose of authorizing progress payments
- Preparing change orders
- Clarifying and interpreting the contract documents
- Reviewing, approving, or taking other appropriate action on submittals such as shop drawings and samples

- Rejecting defective work
- Determining dates of substantial completion and final completion

Conditions of the contract need to be closely coordinated with related documents such as the design and construction agreements and Division 01—General Requirements. Changes in one document may necessitate changes in the other documents.

6.2 General Conditions

General conditions for a project can be any of a number of standard documents published that are applicable to the project delivery method. Provisions of general conditions have broad application and include practices common in the United States. Standard general conditions for construction are prepared and published by professional associations such as the AIA, EJCDC, Design-Build Institute of America (DBIA), and Associated General Contractors (AGC).

Standard general conditions benefit the construction industry by providing documents with a history of many years of use and refinement. The language and provisions of these standard documents have been tested and interpreted by the courts and are well understood and familiar to those concerned. The use of standard general conditions also establishes a baseline from which exceptions, modifications, and additions of the supplementary conditions are clearly identified and evaluated. Writing new general conditions for a project will conceal the unique requirements rather than make them clearly understood.

Standard general conditions are normally based on the concept of a single prime contract for the work. When a project will be constructed under multiple-prime contracts, each contract should use the same general conditions with emphasis on mutual responsibilities. The general conditions should always be bound into the project manual, not simply included by reference, so that they are easily available during progress of the work.

6.2.1 AIA Documents

AIA publishes its "A" Series documents, which contain agreements and general conditions for various project delivery methods, including:

- *AIA Document A295, General Conditions of the Contract for Integrated Project Delivery*
- *AIA Document A201, General Conditions of the Contract for Construction*
- *AIA Document A232/CMa, General Conditions of the Contract for Construction, Construction Manager—Adviser Edition*
- *AIA Document A201/SC, Federal Supplementary Conditions of the Contract for Construction*
- *AIA Document A151, General Conditions of the Contract for Furniture, Furnishings, and Equipment*

- *AIA Document A142, Standard Form of Agreement Between Design-Builder and Contractor*
- *AIA Document A107, Abbreviated Standard Form of Agreement Between Owner and Contractor for Construction Projects of Limited Scope Where the Basis of Payment Is a Stipulated Sum*

AIA publishes other forms of an abbreviated owner-contractor agreement for construction, which contain shortform general conditions to be used for less complex projects.

6.2.2 EJCDC Documents

EJCDC publishes general conditions and agreements for various project delivery methods in conjunction with its engineer-, owner-, and construction-related documents, including:

- *EJCDC C-700, Standard General Conditions of the Construction Contract*
- *EJCDC C-520, Suggested Form of Agreement Between Owner and Contractor on the Basis of a Stipulated Price*
- *EJCDC C-525, Standard Form of Agreement Between Owner and Contractor on the Basis of Cost-Plus*
- *EJCDC D-700, Standard General Conditions of the Contract Between Owner and Design/Builder*
- *EJCDC D-520, Standard Form of Agreement Between Owner and Design-Builder on the Basis of a Stipulated Price*
- *EJCDC D-525, Standard Form of Agreement Between Owner and Design/Builder on the Basis of Cost-Plus*
- *EJCDC E-500, Agreement Between Owner and Engineer for Professional Services*
- *EJCDC P-700, Standard General Conditions for Procurement Contracts*

6.2.3 DBIA Documents

DBIA publishes a Series 500 set of standard documents, including the following:

- *DBIA Document 520, Standard Form of Preliminary Agreement Between Owner and Design-Builder*
- *DBIA Document 525, Standard Form of Agreement Between Owner and Design-Builder—Lump Sum*
- *DBIA Document 530, Standard Form of Agreement Between Owner and Design-Builder—Cost Plus Fee with an Option for a Guaranteed Maximum Price*
- *DBIA Document 535, Standard Form of General Conditions of the Contract Between Owner and Design-Builder*
- *DBIA Document 540, Standard Form of Agreement Between Design-Builder and Designer*
- *DBIA Document 550, Standard Form of Agreement Between Design-Builder and General Contractor—Cost Plus Fee with an Option for a Guaranteed Maximum Price*

6.2.4 Public Agency Documents

Federal, state, and local government agencies usually produce standard general conditions specifically for use on their projects. Some large corporations have also developed their own general conditions for their projects.

When working with unfamiliar general conditions, the A/E must use particular care to adapt the wording of contract documents to the differing provisions to avoid conflicts or omissions. Changes may be required in provisions that the A/E would otherwise treat as standard or routine. Provisions that might be affected include substitutions, submittal procedures, naming of manufacturers, and temporary facilities.

6.3 Common Articles

The *AIA Document A201, General Conditions of the Contract for Construction*; the *EJCDC C-700, Standard General Conditions of the Construction Contract*; and the *DBIA Document 535, Standard Form of General Conditions of the Contract Between Owner and Design-Builder*, address common project concerns. Each document deals with these concerns in a similar manner but with language determined to be in accord with the needs of each organization. In general, the DBIA general conditions provide fewer details and requirements than do the EJCDC and AIA documents because the designer and the builder are acting as a single entity. Some of these common articles deal with:

Work Each of the documents describes work under the basic definitions in the respective first articles. Whereas all define work as the construction and other services required by the contract documents, including labor, material, equipment, and services, AIA Document A201 additionally refers to the contractor's obligation to perform. EJCDC C-700 includes a reference to the documentation required to produce the construction. Both the AIA and the EJCDC general conditions go on to describe the project as the total construction, of which the work under the contract documents may be the whole or part.

Contract Documents

- AIA Document A201 lists the documents that make up the contract documents as the agreement, conditions of the contract, drawings, specifications, addenda issued prior to execution of the contract, other documents listed in the agreement, and modifications issued after execution of the contract and excludes certain procurement documents.
- EJCDC C-700 simply states that the contract documents are those printed or hard-copy documents listed in the agreement, excluding shop drawings, other contractors' submittals, and reports and drawings of subsurface and physical conditions.

Payments The articles dealing with payment procedures are fairly detailed in each of the general conditions. Each sets out requirements for a schedule of values and a method of request for payment and review of application, withholding payment, substantial completion, and final payment. Both AIA Document A201 and EJCDC C-700 define a contract sum.

Termination Each of the general conditions provides detailed procedures for both the owner and the contractor or design-builder to suspend the work without cause or to terminate the contract for cause. In general, the reasons for suspension or termination are similar for each of the general conditions and include failure to make payments, failure to provide adequate workforce or materials, flagrant disregard for laws and regulations, or a breach of the contract documents. The time periods allowed under each general condition vary. In addition, AIA Document A201 allows the contractor to terminate the contract if the work has stopped for 30 consecutive days as a result of a declaration of a national emergency.

Claims and Dispute Resolution Both the AIA and EJCDC general conditions refer most claims or disputes to the A/E first, and a decision is made at this level if possible. AIA Document A201 exempts any claims dealing with hazardous materials, but EJCDC C-700 has no similar exception. For both AIA and EJCDC, the next step after failing resolution with the A/E is to take the dispute to mediation and, failing that, next to arbitration for resolution.

Each of the standard general conditions, though similar in many respects, has developed over time and has evolved into a document that serves the basic requirements of its sponsoring organization. Review these documents closely for each project and determine the source of the conditions in conjunction with the owner and their representative; then coordinate the requirements for the project accordingly. For many of these organizations these documents are always in review, and determining whether any changes have occurred and which set of conditions will be used in a given project will be essential to successful delivery of accurate documentation.

6.4 Supplementary Conditions

The requirements of the general conditions can be modified or expanded in the supplementary conditions to accommodate specific project conditions, owner's requirements, or unusual aspects relating to the roles, rights, duties, and responsibilities of the parties to the contract for construction. Supplementary conditions are not standard for every project and must be prepared based on requirements of the specific project. Some modification of standard general conditions is almost always necessary.

Regardless of the efforts to develop standardized documents, modifications are necessary to customize the requirements to the project. The published guides for preparing supplementary conditions, such as AIA Document A503, give examples of wording for modifications that occur frequently, with a recommended procedure for the preparation of the supplementary conditions. Subjects that require provisions in the supplementary conditions include:

- Insurance requirements
- Progress payments
- Wage rate requirements
- Equal employment opportunity requirements
- Liquidated damages
- Retainage
- Tax-exempt status of the owner

Other items that may be included are expanded definitions and additional rights, duties, responsibilities, or roles to coordinate with other project agreements (e.g., architect/engineer, design-builder, construction manager, etc.).

6.4.1 Preparation of Supplementary Conditions

Deletion or rewording of provisions in standard general conditions should be avoided. The provisions and terminology have been carefully coordinated with companion documents, and the deletion or change of requirements of a provision may affect other documents. When rewording of the general conditions is necessary to adjust for local conditions, owner's requirements, or other legitimate reasons, advice of legal counsel is necessary to avoid conflicts with, or weakening of, unmodified statements elsewhere in the contract documents. Modifications, deletions, and expansions of articles in the general conditions are written in the supplementary conditions in the same order as the related articles occur in the general conditions, and are referenced specifically to the article, paragraph, subparagraph, or clause in the general conditions. Additional articles, paragraphs, subparagraphs, or clauses of supplementary conditions are added by using the next consecutive article, paragraph, subparagraph, or clause number consistent with the provisions of the general conditions. Properly prepared contract documents agree with one another as integral parts of a whole. The documents should agree with one another in terminology and nomenclature. Modifications made to one of the documents should not contradict provisions contained in the other documents. The supplementary conditions must, therefore, be carefully coordinated not only with the general conditions, but also with the other documents, including procurement requirements and the various agreements.

AIA Document A201/SC contains federal supplementary conditions for the standard AIA Document A201 General Conditions; these supplementary conditions are applicable for certain federal government requirements. Additional conditions that may also be required to adapt the documents to project requirements should be provided in addition to AIA Document A201/SC for a specific project.

The practice of printing standard supplementary conditions is not recommended. Supplementary conditions are based on the unique requirements of each specific project. Recommended additions or modifications to the standard general conditions commonly illustrated in published guides are not applicable for every project, and other changes are frequently required. In addition, the supplementary conditions are prepared in conjunction with the procurement requirements because of their interrelationship.

The published guides for preparing supplementary conditions give examples of wording for modifications that occur frequently, with a recommended procedure for the preparation of the supplementary conditions.

6.5 Caution

General and supplementary conditions are conditions of the contract and have legal consequences. The A/E should not prepare general or supplementary conditions without explicit approval and guidance from the owner, the owner's legal counsel, and the owner's insurance adviser.

Chapter 7
General Requirements

7.1 Division 01—General Requirements

The sections in Division 01, which are collectively referred to as the General Requirements, specify administrative requirements, procedural requirements, temporary facilities and controls, performance requirements, and life cycle activities. Administrative and procedural requirements are those relating to the process of contract administration, the assignment of contractual responsibilities, and the methods of communicating, controlling, and assuring quality. Temporary facilities and controls are those put into place for use only during the period of construction and that will be removed when no longer required for construction operations. Performance requirements are related to facility and system performance. Life cycle activities are related to commissioning, facility operation, facility maintenance, and facility decommissioning.

Division 01 sections expand on certain of the administrative and procedural provisions in the conditions of the contract and apply broadly to the execution of the work of all the other sections of the specifications. Administrative and procedural requirements unique to a specific section should be covered in the affected section. Division 01 sections cover general requirements for execution of the work and should be written in language broad enough to apply to sections in Divisions 02 through 49. Without Division 01, these requirements would otherwise be repeated throughout the specifications, and the possibility of conflicts and omissions would be increased. This Division 01 concept adheres to the CSI principle of stating information only once and in the right place. Figure 7.1 illustrates the relationships between Division 01 and each of the other construction documents.

7.2 Organizational Basis for Division 01

The administrative sections of Division 01 parallel the succession of titles in *MasterFormat*® and the procedural sequence of PART 1, PART 2, and PART 3 in *SectionFormat*™ to lay out a consistent and logical organization of titles.

7.2.1 Division 01 and *MasterFormat*®

MasterFormat® provides the major section numbers and titles for Division 01. When applying *MasterFormat*® to Division 01, consider that:

- There are some numbers and titles in *MasterFormat*® that may never appear in a project manual for a construction project, such as the section Operation and Maintenance.

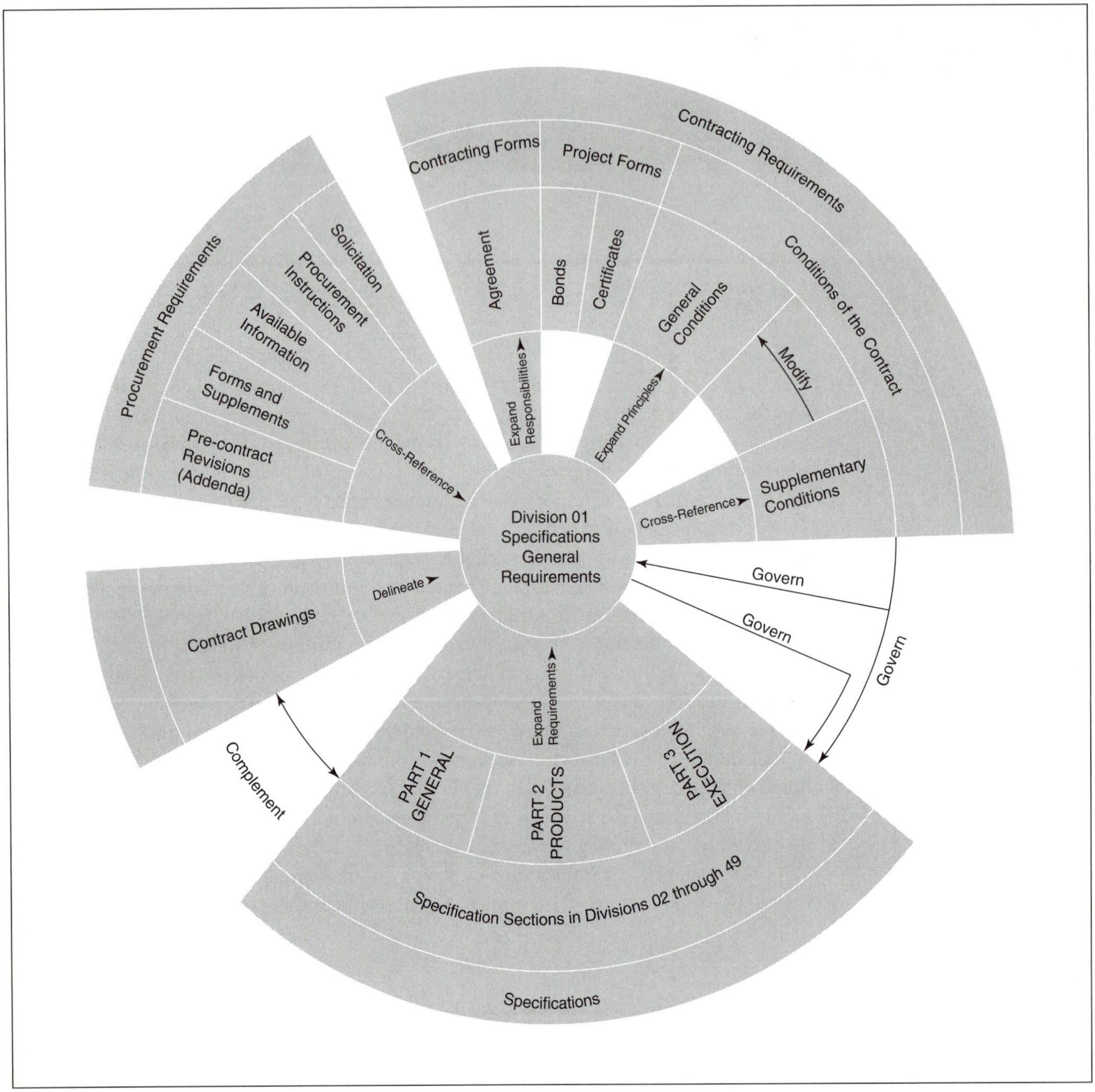

Figure 7.1
Division 01 Relationship to Other Documents

- There are some numbers and titles in *MasterFormat*® with special relevance to project manuals because they deal with activities specifically related to construction activities. The section Work Restrictions and the section Temporary Facilities and Controls are two such examples.
- Because construction specifications are created for service use during the construction stage of the life cycle of a facility, there are numbers and titles in *MasterFormat*® that would be more likely to appear in a stage other than construction, such as facility management.

7.2.2 Division 01 and *SectionFormat™*

Division 01 specifies procedural requirements common to many specification sections and to the project as a whole. Most of these requirements are related to the administrative activities of the project, and others govern products and execution requirements. Accordingly, the numbers and titles of Division 01 in *MasterFormat®* have been arranged to parallel the sequence of information in *SectionFormat™*. Figure 7.2, pages 1 and 2, shows the parallel organization of Division 01 titles to *SectionFormat™*.

DIVISION 01 GENERAL REQUIREMENTS		*SectionFormat™*
01 00 00	**GENERAL REQUIREMENTS**	**PART 1—GENERAL**
01 10 00	**SUMMARY**	**SUMMARY**
	Summary of Work	Section Includes
	Multiple Contract Summary	Products Supplied but Not Installed Under This Section
	Work Restrictions	Products Installed but Not Supplied Under This Section
	Project Utility Sources	Related Sections
01 20 00	**PRICE AND PAYMENT PROCEDURES**	
	Allowances	Allowances
	Unit Prices	Unit Prices
	Alternates	Alternates
	Value Analysis	Measurement Procedures
	Substitution Procedures	Payment Procedures
	Contract Modification Procedures	Alternates
	Payment Procedures	
01 30 00	**ADMINISTRATIVE REQUIREMENTS**	**REFERENCES**
		DEFINITIONS
	Project Management and Coordination	**SYSTEM DESCRIPTION**
	Construction Progress Documentation	Design Requirements, Performance Requirements
	Submittal Procedures	**SUBMITTALS**
	Special Procedures	Product Data
		Shop Drawings
		Samples
		Quality Assurance/Control Submittals
		Closeout Submittals
01 40 00	**QUALITY REQUIREMENTS**	**QUALITY ASSURANCE**
	Regulatory Requirements	Qualifications
	References	Regulatory Requirements
	Quality Assurance	Certifications
	Quality Control	Field Samples
		Mock-ups
		Preinstallation Meetings
01 50 00	**TEMPORARY FACILITIES AND CONTROLS**	
	Temporary Utilities	
	Construction Facilities	
	Temporary Construction	
	Construction Aids	
	Vehicular Access and Parking	
	Temporary Barriers and Enclosures	
	Temporary Controls	
	Project Identification	

(*Continued*)

	DIVISION 01 GENERAL REQUIREMENTS	*SectionFormat™*
01 60 00	**PRODUCT REQUIREMENTS** Common Product Requirements Product Options Owner Furnished Products Product Delivery Requirements Product Storage and Handling Requirements	**DELIVERY, STORAGE, AND HANDLING** Packing, Shipping, Handling, and Unloading Acceptance at Site Storage and Protection Waste Management and Disposal
01 70 00	**EXECUTION AND CLOSEOUT REQUIREMENTS** Examination and Preparation Execution Cleaning and Waste Management Starting and Adjusting Protecting Installed Construction Closeout Procedures Closeout Submittals Demonstration and Training	**PROJECT/SITE CONDITIONS** Project/Site Environmental Requirements Existing Conditions **SEQUENCING** **SCHEDULING** **WARRANTY** Special Warranty
01 80 00	**PERFORMANCE REQUIREMENTS** Facility Performance Requirements Facility Substructure Performance Requirements Facility Shell Performance Requirements Interior Performance Requirements Conveying Equipment Performance Requirements Facility Services Performance Requirements Equipment and Furnishings Performance Requirements Other Facility Construction Performance Requirements Site Construction Performance Requirements	**SYSTEM START-UP** **OWNER'S INSTRUCTIONS**
01 90 00	**LIFE CYCLE ACTIVITIES** Commissioning Facility Operation Facility Maintenance Facility Decommissioning	**COMMISSIONING** **MAINTENANCE** Extra Materials Maintenance Service **PART 2—PRODUCTS** Not Used **PART 3—EXECUTION** Not Used

Figure 7.2
Division 01 Relationship to *SectionFormat™*

- Sections aligning with PART 1—GENERAL are:
  - Summary
  - Price and Payment Procedures
  - Administrative Requirements
  - Quality Requirements
  - Temporary Facilities and Controls
- Sections aligning with PART 2—PRODUCTS are:
  - Product Requirements
  - Performance Requirements
- Sections aligning with PART 3—EXECUTION are:
  - Execution and Closeout Requirements
  - Life Cycle Activities

The relationship between Division 01 and *SectionFormat™* defines an organizing principle. The logic of this organization provides the flexibility by combining procedural requirements into a few sections or dividing them into many as dictated by the complexity of the project.

7.3 Division 01 Relation to Other Documents

Each of the contract documents complements one another and should not duplicate specific information. The provisions of the conditions of the contract and the agreement apply broadly to the work of the project, and Division 01 expands on these provisions, giving just enough detail to apply broadly to the other specification sections. The specifications and drawings further expand on Division 01 and provide detailed requirements for specific portions of the work.

7.3.1 Procurement Requirements

By definition in American Institute of Architects (AIA) and Engineers Joint Contract Documents Committee (EJCDC) general conditions, procurement requirements are not a part of the contract documents. Provisions that are stated only in procurement requirements are not enforceable during administration of the construction contract.

The procurement requirements should not repeat Division 01 provisions but should instead refer to the appropriate Division 01 sections, by number and title, to direct the bidder to relevant information.

Provisions applicable only during the bidding stage, such as document acquisition and cost, bid security requirements, and bid opening dates, should be stated in the procurement requirements rather than in Division 01.

7.3.2 Owner-Contractor Agreement

The owner-contractor agreement includes by reference all other contract documents. Certain provisions of the owner-contractor agreement are supplemented by the provisions of Division 01 sections that define in greater detail the responsibilities of the parties to the contract.

For example, Section 01 10 00—Summary specifies administrative details and the responsibilities of the owner and the contractor, including a summary of the work, identification of other contracts, work restrictions such as limitations on access and use of the site, and the need to accommodate the owner's occupancy during construction.

Similarly, Section 01 20 00—Price and Payment Procedures expands on the related provisions stated in the agreement, such as allowances, alternates, and unit prices.

7.3.3 General Conditions

The general conditions of the contract govern the execution of the work and apply broadly to sections of the specifications, including those in Division 01. Among the topics usually covered are property surveys, temporary utilities and services, warranties, progress schedules, record documents, submittals, cutting and patching, cleaning, schedule of values, applications for payment, and closeout procedures. As these topics are covered only in very broad terms, they must be specified in further detail in Division 01 to tailor requirements to a specific project. Although the topics are addressed in two locations, the Construction Specifications Institute (CSI) principle of "say it once" is not violated because the requirements are stated in two different levels of detail, without duplication or conflict. For example, the general conditions broadly state that the owner may employ separate contractors and will provide coordination.

If separate contractors are in fact employed, Division 01 could specify this in Section 01 12 00—Multiple Contract Summary and could further identify the contracts, the scope of work in each, and the responsibilities of the owner, the architect/engineer (A/E), and each contractor in the coordination process.

7.3.4 Supplementary Conditions

Supplementary conditions should not repeat information contained in the General Conditions or encroach on topics that belong more appropriately in Division 01. Proper coordination between Division 01 and supplementary conditions will avoid conflicts, omissions, and duplications.

Other than a statement establishing the authority of Division 01, the supplementary conditions should not contain general references to Division 01. Statements in the supplementary conditions referring to specific sections in Division 01 are not necessary but may be helpful to the user.

For example, a statement may be added to the supplementary conditions stating that shop drawings, product data, samples, and other submittals shall be prepared and submitted according to the requirements of specification Section 01 33 00—Submittal Procedures.

7.3.5 Other Specification Sections

Specification sections are subject to the administrative and procedural requirements of the conditions of the contract and of Division 01. Project requirements become more specific for each successive level of a three-tier hierarchy:

- The general conditions state provisions in broad terms.

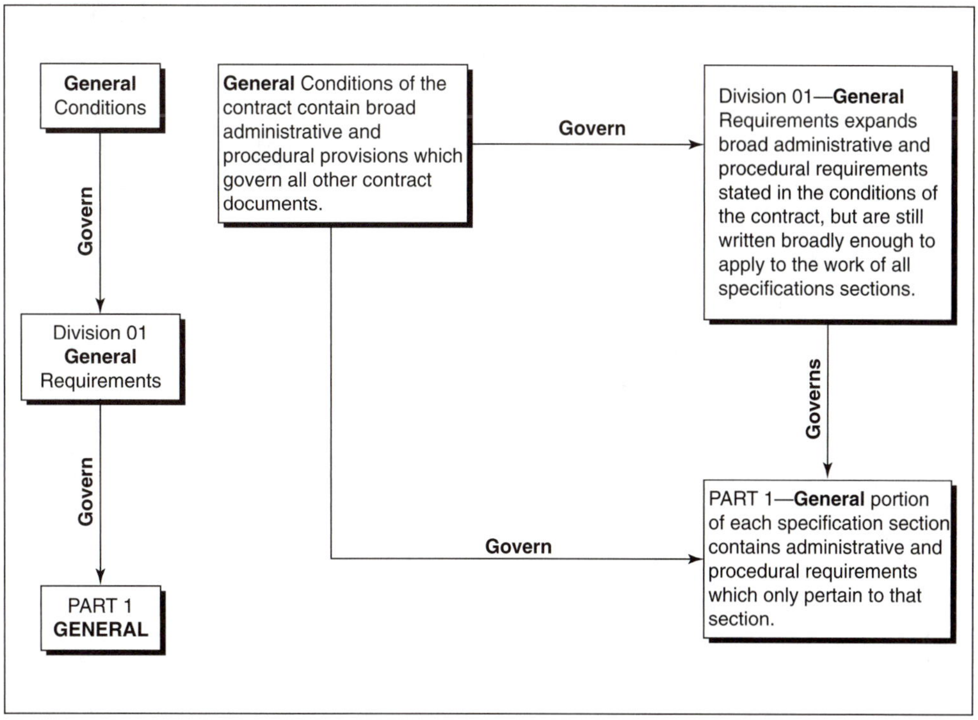

Figure 7.3
Hierarchy of General Administrative and Procedural Requirements

- Division 01—General Requirements elaborate on the broad provisions of the conditions of the contract, but it is still written broadly enough to apply to the work of all specification sections.
- PART 1—GENERAL in a specification section becomes very specific in project requirements for that section only.

Refer to Figure 7.3 for illustration of this hierarchy. For example, the general conditions establish the fact that testing may be required. Division 01 specifies the broad administrative and procedural requirements for testing laboratory services such as who selects the laboratory, who pays for which services, qualifications of the testing laboratory, submittal of reports, and the contractor's responsibilities for cooperation. The individual specification sections state further requirements of particular tests to be performed, the number of tests required, and other details applicable only to the work or products in that section. As another example, although routine submittal procedures are covered in Division 01, the specific items for which shop drawings or samples must be submitted are unique to individual specification sections, and specific requirements must be identified in those sections. Figure 7.4 gives examples of the levels of detail of general administrative and procedural requirements found in various construction documents.

Although Division 01 requirements are generally arranged in the order of the three parts of *SectionFormat™*, PART 1—GENERAL should, in most cases, contain the bulk of these provisions.

AIA Document A201, General Conditions of the Contract for Construction

3.12 SHOP DRAWINGS, PRODUCT DATA AND SAMPLES

3.12.1 Shop Drawings are drawings, diagrams, schedules and other data specially prepared for the Work by the Contractor or a Subcontractor, Sub-subcontractor, manufacturer, supplier or distributor to illustrate some portion of the Work.

3.12.2 Product Data are illustrations, standard schedules, performance charts, instructions, brochures, diagrams and other information furnished by the Contractor to illustrate materials or equipment for some portion of the Work.

3.12.3 Sample are physical examples that illustrate materials, equipment or workmanship and establish standards by which the Work will be judged.

3.12.4 Shop Drawings, Product Data, Samples and similar submittals are not Contract Documents. Their purpose is to demonstrate the way by which the Contractor proposes to conform to the information given and the design concept expressed in the Contract Documents for those portions of the Work for which the Contract Documents require submittals. Review by the Architect is subject to the limitations of Section 4.2.7. Informational submittals upon which the Architect is not expected to take responsive action may be so identified in the Contract Documents. Submittals that are not required by the Contract Documents may be returned by the Architect without action.

3.12.5 The Contractor shall review for compliance with the Contract Documents, approve and submit to the Architect Shop Drawings, Product Data, Samples and similar submittals required by the Contract Documents in accordance with the submittal schedule approved by the Architect or in the absence of an approved submittal schedule, with reasonable promptness and in such sequence as to cause no delay in the Work or in the activities of the Owner or of separate contractors.

3.12.6 By submitting Shop Drawings, Product Data, Samples and similar submittals, the Contractor represents to the Owner and Architect that the Contractor has (1) reviewed and approved them, (2) determined and verified materials, field measurements and field construction criteria related thereto, or will do so and (3) checked and coordinated the information contained within such submittals with the requirements of the Work and of the Contract Documents.

3.12.7 The Contractor shall perform no portion of the Work for which the Contract Documents require submittal and review of Shop Drawings, Product Data, Samples or similar submittals until the respective submittal has been approved by the Architect.

3.12.8 The Work shall be in accordance with approved submittals except that the Contractor shall not be relieved of responsibility of deviations from requirements of the Contract Documents by the Architect's approval of Shop Drawings, Product Data, Samples or similar submittals unless the Contractor has specifically informed the Architect in writing of such deviation at the time of submittal and (1) the Architect has given written approval to the specific deviation as a minor change in the Work, or (2) a Change Order or Construction Change Directive has been issued authorizing the deviation. The Contractor shall not be relieved of responsibility for errors or omissions in Shop Drawings, Product Data, Samples or similar submittals by the Architect's approval thereof.

3.12.9 The Contractor shall direct specific attention, in writing or on resubmitted Shop Drawings, Product Data, Samples or similar submittals, to revisions other than those requested by the Architect on previous submittals. In the absence of such written notice, the Architect's approval of a resubmission shall not apply to such revisions.

3.12.10 The Contractor shall not be required to provide professional services that constitute the practice of architecture or engineering unless such services are specifically required by the Contract Documents for a portion of the Work or unless the Contractor needs to provide such services in order to carry out the Contractor's responsibilities for construction means, methods, techniques, sequences and procedures. The Contractor shall not be required to provide professional services in violation of applicable law. If professional design services or certifications by a design professional related to systems, materials or equipment are specifically required of the Contractor by the Contract Documents, the Owner and the Architect will specify all performance and design criteria that such services must satisfy. The Contractor shall cause such services or certifications to be provided by a properly licensed design professional, whose signature and seal shall appear on all drawings, calculations, specifications, certifications, Shop Drawings and other submittals prepared by such professional. Shop Drawings and other submittals related to the Work designed or certified by such professional, if prepared by others, shall bear such professional's written approval when submitted to the Architect. The Owner and the Architect shall be entitled to rely upon the adequacy, accuracy and completeness of the services, certifications and approvals performed or provided by such design professionals, provided the Owner and Architect have specified to the Contractor all performance and design criteria that such services must satisfy. Pursuant to this Section 3.12.10, the Architect will review, approve or take other appropriate action on submittals only for the limited purpose of checking for conformance with information given and the design concept expressed in the Contract Documents. The Contractor shall not be responsible for the adequacy of the performance and design criteria specified in the Contract Documents.

Supplementary Conditions

3. In Paragraph 3.12.5 add the following: "Prepare and submit Shop Drawings, Samples and other data in accordance of Section 01 33 00—Submittal Procedures."

Section 01 33 00- Submittal Procedures, PART 1—GENERAL

1.03 SUBMITTAL PROCEDURES

A. Transmit each submittal with AIA form G810
B. Sequentially number transmittal forms. Resubmittals to have original number with an alphabetic suffix.
C. Identify Project, Contractor, Subcontractor or supplier: pertinent Drawing sheet and detail number and specification Section number as appropriate.
D. Affix Contractor's stamp, signed or initialed certifying that review verification or product's required field dimensions, adjacent construction Work, and coordination of information, is in accordance with the requirements of the Work and Contract Documents.
E. Schedule submittals to expedite the Word and deliver to Architect/ Engineer at business address. Coordinate submittal of related items.
F. Identify variations from Contract Documents and product or system limitations which may be detrimental to successful performance or the completed Work.
G. Provide space for Contractor and Architect/ Engineer review stamps.
H. Revise and resubmit submittals as required, identify all changes made since previous submittal.
I. Distribute copies of reviewed submittals to concerned parties. Instruct parties to promptly report any inability to comply with provisions.

Section 05 42 00—Cold-Formed Metal Joist Framing

1.05 SUBMITTALS

A. Submit shop drawings and product data.
B. Indicate on shop drawings component details, framed openings, welds, type and location of fasteners, and accessories or items required of other related work.

Section 09 60 00—Access Flooring

1.09 SUBMITTALS

A. Submit shop drawings and product data.
B. Indicate flooring layout, interruptions to grid, special sized panels required, panels drilled or cut-out for services, appurtenances or interruptions, edge details, elevation differences, grilles, and registers.

Figure 7.4

AIA Document A201. Level of Detail for General Administrative and Procedural Requirements

PART 2—PRODUCTS and PART 3—EXECUTION may also include some items that relate to administrative and procedural requirements. For example, quality control requirements covered in Division 01 may include general erection and approval procedures for the use of mock-ups as quality standards.

PART 1—GENERAL of the section on glazed curtain walls may stipulate that a mock-up be tested for air and water infiltration according to the requirements of referenced American Society for Testing and Materials International (ASTM) or American Architectural Manufacturers Association (AAMA) standards.

PART 2—PRODUCTS may require the manufacturer to perform the tests as part of a source quality control program under supervision of the owner's testing laboratory consultants.

PART 3—EXECUTION might require the manufacturer to provide factory representatives to ensure that field erection and quality control procedures conform to those used in the approved mock-up. Although the Division 01 role in governing the work has been accepted in practice for many years, this authority is not explicitly stated in either the AIA or EJCDC general conditions. Until that change is made, the authority should be established by a provision in the supplementary conditions as follows:

> *Sections of Division 01—General Requirements govern the work of all sections of the specifications.*

A similar statement should be included in the supplementary conditions of Division 01 for projects using federal, state, city, or county documents that do not already grant such authority.

7.3.6 Drawings

Drawings graphically define certain Division 01 subject matter, such as the extent of work, and may define related work that may or may not be part of the contract.

Purely procedural sections of Division 01, such as submittals, quality requirements, and closeout procedures, need no coordination with the drawings.

Other sections, such as those covering use of the site, phased construction, and multiple prime contracts, usually require extensive delineation on the drawings and close coordination between the drawings and Division 01.

7.4 Writing Division 01 Sections

Division 01 sections are organized and written in the same outline style and three-part *SectionFormat™* as sections in Divisions 02 through 49. Section 01 57 00—Temporary Controls, for example, might include not only the PART 1 administrative and procedural provisions for surface water control, but also PART 2 requirements for containment devices and PART 3 requirements for their installation and subsequent removal. Many of the Division 01 sections, however, will have no PART 2—PRODUCTS or PART 3—EXECUTION when limited strictly to administrative and procedural requirements. Whenever this is the case, the unused parts are listed but noted as "not used" (e.g., PART 2—PRODUCTS—Not Used).

7.4.1 Coordination with Division 01—General Requirements

A number of articles in the general conditions are usually expanded in Division 01—General Requirements to accommodate specific administrative and procedural requirements of a project. Articles on allowances, progress schedules, record documents, shop drawings, product data and samples, schedule of values, and closeout procedures and submittals are commonly developed in greater detail in Division 01. General and supplementary conditions are written in a paragraph style, while Division 01 is written in the three-part *SectionFormat™*.

Proper use of Division 01 avoids mixing conditions of the contract with the administrative and procedural details of the specifications. A comparison of conditions of the contract and Division 01—General Requirements is shown in Figure 7.5. Coordination is necessary between each section of Division 01 and the sections of Divisions 02 through 49, and among the sections of Division 01 that cover interrelated administrative and procedural requirements. For example, Section 01 14 00—Work Restrictions contains restrictions on the operations of the contractor based on continuing occupancy by the owner that must be maintained. At the same time, Section 01 35 00—Special Procedures may establish similar restrictions based on the type of activity or the type of project. The specifier should verify that each requirement is in the one location offering the basis for imposing the requirement.

CONDITIONS OF THE CONTRACT	DIVISION 01-GENERAL REQUIREMENTS
Are inherent part of the Agreement	An inherent part of the Specifications
With the Agreement govern the content of the entire contract	Administratively governs the Specification sections
Contain contractual principles applicable to most projects with supplements for a particular project	Contains specifics directly applicable to a particular project

GENERAL CONDITIONS	SUPPLEMENTARY CONDITIONS	SECTIONS OF DIVISION 01
Are broad contractual conditions	Modify the contractual conditions	Contain specific administrative and procedural requirements
Contain the constants	Modify the constants for a specific region or project	Contain variables directly applicable for specific project
Relatively static content thus allowing the use of published standard documents	Take precedence over general conditions	Must be written separately for each project
	Must be written separately for each project	
ARE NOT SPECIFICATIONS		ARE SPECIFICATIONS

Figure 7.5
Conditions of the Contract versus Division 01—General Requirements

CONDITIONS OF THE CONTACT
Supplementary Conditions

Delete Paragraph 11.8—Cash Allowances, and substitute the following:

11.8—Allowances: Include in the Contract Sum all allowances stated in the Contract Documents. Allowances include specific monetary sums and quantities of work for certain scheduled items. Refer to Section 01 21 13—Cash Allowances and Section 01 21 19—Quantity Allowances for allowance descriptions and requirements.

SPECIFICATIONS
Division 01—General Requirements

SECTION 01 21 13
CASH ALLOWANCES

1.06 SCHEDULE OF CASH ALLOWANCES

A. Section 04 20 00—Unit Masonry: Include the unit price of $350.00 per thousand for purchase and delivery of facing brick. Stipulate the estimated quantity on the Bid Form. Include installation costs in Contract Sum.

Division 04—Masonry

SECTION 04 20 00
UNIT MASONRY

1.05 ALLOWANCES

A. Provide selected facing brick under cash allowance specified in Section 01 21 13—Cash Allowances.

SECTION 01 21 19
QUANTITY ALLOWANCES

1.06 SCHEDULE OF QUANTITY ALLOWANCES

A. Section 09 68 00—Carpeting: Provide 200 SY, including purchase, delivery and installation of Type "A" Sheet Carpet.

B. Section 09 68 00—Carpeting: Provide 400 SY, including purchase, delivery and installation of Type "B" Carpet Tile.

Division 09—Finishes

SECTION 09 68 00
CARPETING

1.05 ALLOWANCES

A. Provide carpet quantities specified under Section 01 21 19—Quantity Allowances.

Figure 7.6
Sample Use of Allowances

Division 01 sections on allowances and alternates require particular coordination with the sections in other divisions that specify the affected products (refer to Figure 7.6). Division 01 sections that cover cash allowances for materials should include procedural details for the selection of specific products, identify the allowance items by item number and title, along with the type of product and the dollar amount to be allowed, and cross-reference the affected sections in Divisions 02 through 49. In turn, the sections specifying the installation of products that will be purchased under a cash allowance should cross-reference the Division 01 section specifying the dollar amounts, without repeating the Division 01 administrative and procedural requirements.

Section 01 23 00—Alternates should identify the alternates by number and give the procedures for bidding and for consideration and acceptance of alternates. It should also describe the changes to other work required by acceptance of each alternate and list the sections affected by each. Sections specifying products affected by an alternate should cross-reference Section 01 23 00—Alternates.Work involving rehabilitation, owner-occupied projects, hazardous materials, phased and fast-track techniques, construction management, multiple-prime contracts, purchasing, and overall performance requirements requires special attention to Division 01 to carefully describe the administrative and procedural responsibilities of each of the participants involved in the construction process.

As many of the contractual relationships and responsibilities described in Division 01 sections for any project are required by the owner, the A/E should obtain instructions for preparing the affected sections. The individuals involved in the preparation of a project manual represent various professional disciplines and may be members of the same firm or consultants under contract to the A/E. The A/E should establish the basic responsibilities of the team members for preparation of the Division 01 requirements and for overall coordination of the specifications. This involves close coordination of Division 01 provisions at an early stage in the production of the contract documents to avoid conflicts, duplication, and omissions in the project manual. The coordination effort should include:

- Distributing a draft of Division 01 to the contributors to the specifications early in project development
- Preparing of a list by the consultants and other contributors of administrative, procedural, and temporary facility and control requirements for specific sections or divisions
- Comparing this list of requirements with the draft of Division 01
- Determining which of these requirements should be covered in Division 01 and which are specific only to individual specification sections
- Noting overlapping requirements and omissions and resolving conflicts
- Revising the Division 01 draft for review by specification contributors

Multiple sections within an individual division have administrative and procedural requirements unique to that division. However, the common requirements should be specified in Division 01. Often, some sections or divisions are written by different professional disciplines. To ensure proper coordination, the general requirements common to sections or divisions should be in their proper location in Division 01. This can be accomplished either by distributing drafts of the applicable Division 01 sections to each of the team members for markup or by developing a checklist to allow the team members to note their requirements.

Administrative and procedural requirements unique to an individual section should be covered in the affected sections.

The practice of including Basic Requirements or Special Provisions sections in individual divisions is not recommended and may result in conflicts with Division 01.

7.5 Commonly Used Division 01 Sections

Several of the more commonly used Division 01 sections are listed in the following paragraphs with some guidelines for the general types of information that might be included in each. For detailed examples of these sections, refer to master guide specification systems such as the AIA/ARCOM (Architectural Computer Services, Inc.) MASTERSPEC®, Construction Sciences Research Foundation (CSRF) SPECTEXT®, or Building Systems Design (BSD) SpecLink®.

7.5.1 Section 01 10 00—Summary

Section 01 11 00—Summary of Work Use this section to describe the work if all of the work required by the project manual and drawings is to be accomplished under one contract. If the section includes a description of the project, specify the type of contract used, and fully and accurately define the extent of the contract. To further define the scope of the section, include owner-furnished material and equipment installed by the contractor. The A/E should identify work under other contracts and specify required work sequences if phasing of the work is critical.

Section 01 12 00—Multiple Contract Summary Use this section if the work described in the project manual and drawings is to be accomplished by more than one contract. The situation of multiple contracts may arise from techniques including multiple-prime contracts, construction management, and, in some cases, design-build. If the section includes a description of the project, specify the type of contract used, and fully and accurately define the extent of the contract. To further define the scope of the section, include owner-furnished material and equipment installed by which contractor.

Section 01 14 00—Work Restrictions Specify owner requirements for early partial occupancy of new construction or occupancy of existing buildings. Carefully define restrictions on the contractor's use of premises, especially if the project area will be occupied during construction. Limit this section to restrictions on the contractor's use of the premises, placing protection for the owner's occupancy in Section 01 56 00—Temporary Barriers and Enclosures and placing requirements to provide access in Section 01 53 00—Temporary Construction and Section 01 55 00—Vehicular Access and Parking.

- Obtain and include occupancy and access requirements from the owner.
- Determine the number of days of prior notice required for work in occupied areas.
- Specify work areas and access routes to be used by the contractor.
- State work and utility outage schedules imposed as a result of the owner's occupancy of affected areas and any owner-required working hours for the contractor's operations.

7.5.2 Section 01 20 00—Price and Payment Procedures

Section 01 21 00—Allowances Allowances may be used when certain design decisions have not been made and when quantities or unknowns cannot be determined prior to the start of work. Allowances are typically cash allowances or quantity allowances. Include only if directed by the owner and in an amount determined by the owner.

Section 01 22 00—Unit Prices Always include the procedures regarding unit prices and measurement for payment. Define who is responsible for actual measurement of unit price quantities, including procedures for administration and coordination. There should also be a reference to the unit price schedule as incorporated into the agreement. The preferred practice is to include a listing of unit price items showing the item number, a designation of the item, and the unit and method of measurement in the unit price schedule incorporated into the agreement. Detailed discussion of aspects of the work associated with each unit price item should be included in the relevant specification sections.

Section 01 23 00—Alternates Obtain in writing from each team member a list of alternates included in their work and the effect of other alternates on their work. Clearly define the limits of the work of each alternate and the procedure that will be used for evaluation and acceptance. Refer to each specification section that contains technical requirements for alternate products or systems or their installation. Where possible, refer to other sections covering work that is significantly affected by alternates. Refer to drawings when alternates affect extent of work.

Section 01 24 00—Value Analysis Specify general requirements and procedures for conducting value analysis, including description of participants, schedule, costs, evaluation, and reports. Indicate the method of acceptance and the implementation of results.

Section 01 25 00—Substitution Procedures Specify basic requirements for consideration of the contractor's proposals for substitution of products, including general rules and procedures for determining whether products are equivalent. If substitutions are allowed both during bidding or negotiating and after the contract is awarded, the procedures here may serve for both time frames, but the procurement requirements will need to refer to this section as the basis for the process during the bidding or negotiating. Specify requirements and procedures for substitution of a specified installer.

Section 01 26 00—Contract Modification: Procedures Provide a method to allow the contractor or the owner to request clarification about an item of work insufficiently described or detailed in the contract documents. The method should allow the A/E to respond with a clarification that does not require a change to the contract sum or time.

A standardized request for interpretation (RFI) form with space for requesting an interpretation and space for a response will aid the A/E in answering questions. A copy of the form should be placed in the project manual and distributed to the contractor for use. Refer to Figure 7.7 for a sample RFI form.

Specify the modification procedures to be followed and the data required to process contract modifications and to substantiate claims for extra time and costs. Stipulate the types of forms to be used by the contractor and the A/E to request prices for changes and to document the type of modifications to the contract.

Section 01 29 00—Payment Procedures Specify related requirements for the contractor's submittal of a schedule of values. Specify requirements for the contractor's submittal of applications for payment. Identify specific steps in each process. Identify forms to be used, how they are to be executed, how many copies are required, and when they are to be submitted.

7.5.3 Section 01 30 00—Administrative Requirements

Section 01 31 00—Project Management and Coordination Specify requirements and procedures for project coordination, project meetings, and project/site administration. Specify requirements for coordination drawings. If multiple-prime contracts are being used, assign coordination responsibilities to the appropriate

Knowledge for Creating and Sustaining the Built Environment

REQUEST FOR INTERPRETATION

Project:	**The Money Bank**	R.F.I. Number:	**006**
	Dollar, USA	Form:	**Erectum Strait Builders**
To:	**Design Dreamers**	Date:	**June 14, 2010**
	Form and Function, USA	A/E Project Number:	**03-123.00**
Re:	**Door and Teller Area**	Contract For:	**General Construction of Bank**

Specification Section: Paragraph: Drawing Reference: A-101/A-601 Detail:

Request:

The first floor plan indicates only one door to the tellers area. The door schedule has two doors with teller area identifiers. Which is correct?

Signed by: Date: **June 14, 2010**

Response:

Only one door is required as located per floor plan.

☒ Attachments

Response From: **Design Dreamers** To: **Erectum Strait Builders** Date Rec'd: **June 16, 2010** Date Ret'd: **June 19, 2010**

Signed by: Date: **June 19, 2010**

Copies: ☐ Owner ☐ Consultants ☐ ___ ☐ ___ ☐ ___ ☐ ___ ☒ File

Page 1 of 1

July 1994
CSI Form 13.2A

Figure 7.7
CSI Form 13.2A. Request for Interpretation

contractor. Specify requirements for supervisory personnel. Regardless of the prime contract arrangement, use this section to specify requirements regarding the administration of subcontractors and coordination with other contractors and the owner. When applicable, address requirements for preconstruction meetings, preinstallation meetings, coordination meetings, and progress meetings. If the project is constructed under multiple-prime contracts, determine responsibility for project meetings, whether one of the contractors or the construction manager. Specify frequency and location of meetings.

Section 01 32 00—Construction Progress Documentation Specify requirements for the contractor's submittal of construction progress schedule, including type of schedule and time allowed for submittal. Specify requirements for daily construction reports. Specify type of photographic documentation, whether print media, digital still, or video.

Section 01 33 00—Submittal Procedures Specify requirements for a schedule of submittals. Specify requirements for shop drawings, product data, and samples, including reports and data to be processed as a basis prior to accepting the product or activity. Specify shop drawing format (i.e., prints, reproducible, or electronic deliverables), required number of copies to be submitted, and required number of copies to be retained for the owner's use.

Section 01 35 00—Special Procedures Specify procedures for projects requiring exceptional or unusual workmanship of work including restorations, preservations, historic restorations, alterations, and hazardous material abatement. Specify procedures for specific facility types, including airports, detention facilities, hospitals, shopping centers, and nuclear facilities.

7.5.4 Section 01 40 00—Quality Requirements

Specify the procedures to control, measure, and report the quality of the construction and performance of the contractor. Specify the construction, review, and use of mock-ups and field samples applied or assembled at the site as quality standards. Specify qualifications of testing laboratory, limitations of authority, requirements for contractor cooperation, responsibilities for payment, and reference to required tests.

7.5.5 Section 01 50 00—Temporary Facilities and Controls

All team members should review this section and any subordinate sections. If the project is constructed under multiple-prime contracts, assign responsibilities for temporary facilities to the appropriate contractors. In renovation work, rehabilitation projects, or additions to existing facilities, specify whether existing utilities, services, or facilities are available for construction use and whether existing elevators and toilets may be used by construction personnel.

Section 01 51 00—Temporary Utilities Include temporary water, fire protection, electric service, lighting, telephone, and heating, cooling, and ventilating. Specify responsibility for installation and usage costs. Provide for equipment having special requirements.

Section 01 52 00—Construction Facilities For field offices and sanitation facilities, identify the contractor responsible for installation and costs. Specify requirements and limitations for location.

Section 01 53 00—Temporary Construction Specify requirements for temporary structures required to accommodate access to the construction or to accommodate the owner's operations. Limit the specifications in this section to the structural components. Facilities for temporary access and parking are located in Section 01 55 00—Vehicular Access and Parking.

Section 01 54 00—Construction Aids Specify requirements for temporary elevators, hoists, cranes, scaffolding and platforms, and swing staging. Include requirements for using tools and equipment during construction. Indicate requirements for sharing construction aids when multiple contracts are used.

Section 01 55 00—Vehicular Access and Parking Specify requirements for temporary facilities required to accommodate access and parking for the construction or to accommodate the owner's operations. Specify the procedures related to maintaining and controlling traffic and access to the construction. Limit these specifications to maintaining and controlling access; special restrictions on the access of the contractor are specified in Section 01 14 00—Work Restrictions.

Section 01 56 00—Temporary Barriers and Enclosures Include requirements for facilities and procedures for protecting facility users and existing spaces during construction. Specify requirements for fences, noise and dust barriers, and removal of hazardous waste generated as a result of construction operations.

Section 01 57 00—Temporary Controls Include administrative and procedural provisions for surface water control and containment devices and their installation and subsequent removal. Specify erosion and sediment control. Include pest control requirements.

Section 01 58 00—Project Identification Specify requirements and limitations for project identification and temporary signs.

7.5.6 Section 01 60 00—Product Requirements

Section 01 62 00—Product Options Specify the basic requirements for contractors' options in selecting products, including general rules and procedures for determining whether products are equivalent.

Section 01 64 00—Owner-Furnished Products This section may be used to define the administrative requirements for owner-furnished products. As appropriate, state basic characteristics of products furnished by the owner for installation by the contractor. Specify basic requirements and procedures for coordinating, scheduling, and receiving products to ensure that the condition of the products at the time of delivery is acceptable. Extraordinary precautions beyond the requirements of Section 01 65 00—Product Delivery Requirements can also be included. If this section is used, the basic list of the products that are owner furnished should be located here and cross-referenced in Section 01 10 00—Summary.

Section 01 65 00—Product Delivery Requirements Specify the basic requirements for packing, shipping, delivery, and acceptance of products at the site.

Section 01 66 00—Product Storage and Handling Requirements Specify the basic requirements for handling and storage of products at the site.

7.5.7 Section 01 70 00—Execution and Closeout Requirements

Section 01 71 00—Examination and Preparation Specify extraordinary requirements for preparing for construction, including initial construction layout, protection of adjacent construction, and surveying. Specify the basic requirements for preparing to install, erect, or apply products.

Section 01 73 00—Execution Specify the basic requirements for installing, applying, or erecting new and prepurchased products. Specify the basic requirements for installing owner-furnished products, but refer to Section 01 64 00—Owner-Furnished Products for the product characteristics. Specify procedures for dealing with existing material and equipment, making distinctions among several possibilities:

- Material and equipment to be salvaged (removal, repair, and storage) and reinstalled in the work
- Material and equipment to be salvaged and retained by the contractor with the salvage value accruing to the owner
- Material and equipment to be salvaged and delivered to the owner

Section 01 74 00—Cleaning and Waste Management Specify requirements for cleaning during construction and final cleaning. Include requirements for recycling materials.

Section 01 75 00—Starting and Adjusting Specify schedule for starting and adjusting equipment and systems. Describe requirements for starting, adjusting, and servicing equipment. Include requirements for seasonal operation. Include method for reporting compliance with specified requirements.

Section 01 76 00—Protecting Installed Construction Include basic requirements and procedures for protecting finished work and existing finishes, if applicable, while continuing construction. Describe remedial and replacement requirements.

Section 01 77 00—Closeout Procedures Specify administrative procedures for substantial completion and final completion of the work. Describe the procedures required to identify, correct, and complete punch-list items.

Section 01 78 00—Closeout Submittals Specify requirements for closeout submittals such as record documents, warranties, spare parts, and maintenance and operation data. Specify number and form of data to be submitted, such as manuals.

Section 01 79 00—Demonstration and Training Specify requirements and procedures for the demonstration of the products and systems used in the facility, including training of the owner's operations and maintenance staff.

7.5.8 Section 01 80 00—Performance Requirements

Section 01 81 00—Facility Performance Requirements Specify general performance applicable to the entire facility such as sustainability, energy, and environmental design requirements.

- U.S. Green Building Council (USGBC)–Leadership in Energy and Environmental Design (LEED®) buildings registered for certification would include project-specific requirements and goals with submittal instructions to be used during the construction phase of the project.
- Other sustainable design goals and systems for buildings would also be included in these sections.

Sections 01 82 00 through 01 89 00 Specify performance requirements of the facility that are specified in multiple specification sections, including substructure, superstructure, interiors conveying equipment services, equipment and furnishings, site, and other locations that may be project specific.

7.5.9 Section 01 90 00—Life Cycle Activities

Section 01 91 00—Commissioning Specify general requirements for the commissioning of the facility and its systems, including the specific responsibilities of the commissioning agent and other contractors. Specify requirements for the evaluation of the system performance. Specify procedures for testing, adjusting, and balancing facility systems as part of the commissioning process. Coordinate with Section 01 75 00 for starting and adjusting equipment prior to commissioning.

Section 01 92 00—Facility Operation Describe requirements and procedures for operating the facility as a part of a contract.

Section 01 93 00—Facility Maintenance Describe requirements and procedures for maintenance of the facilities, including provisions for maintenance records and logbooks. Specify procedures for recycling programs.

Section 01 94 00—Facility Decommissioning Specify basic requirements for deactivating or removing a facility or its systems or equipment from operation, including protection of deactivated facilities, deconstruction/demolition, and removal. Specify procedures related to abatement, removal, and disposal of hazardous materials.

7.5.10 Cautions

Division 01 administrative and procedural requirements and temporary facilities and controls involve both direct and indirect costs. Overspecifying in Division 01 may increase construction costs unnecessarily and may create excessive obligations for the entity administering the construction contract.

7.6 Specifying Allowances

The use of allowances requires written provisions in several of the procurement documents that must be carefully coordinated with each other. Typically, allowance requirements will appear in the procurement requirements, conditions of the contract, Division 01—General Requirements, and the individual specification sections for the items covered by the allowances. The documents must state:

- Exactly what is included under the allowance
- Who is to authorize items covered by allowances
- How costs will be adjusted if the actual price, quantity, or time varies from the amount stated

7.6.1 Conditions of the Contract

Both AIA and EJCDC publish standard general conditions containing articles pertaining to cash allowances that can be summarized as follows:

The Allowances article in *AIA Document A201, General Conditions of the Contract for Construction*, provides for cash allowances that are included in the contract sum. However, the provisions limit the cash allowance to materials and equipment delivered to the site, including taxes. Items must be furnished by persons designated by the owner; however, the contractor may refuse to contract with them if the contractor provides a reasonable objection. The contractor's costs for labor, unloading and handling at the site, installation, overhead, profit, and other expenses are included in the contract sum unless otherwise provided in the contract documents. The article also states that the contract sum will be adjusted by change order to reflect the cost difference of the allowance.

The Cost of the Work; Allowances; Unit Price of Work article in *EJCDC C-700, Standard General Conditions of the Construction Contract*, provides for cash allowances that are included in the contract sum. The cash allowances are limited by the article's provisions to materials and equipment delivered to the site, including taxes. Work must be done by subcontractors and suppliers acceptable to the owner and engineer. The paragraph also states that prior to final payment, an appropriate change order will be issued to reflect the actual cost and adjustment of the contract price. If requirements stated in the general conditions must be modified to include other items, such as installation, quantities, or time, then the appropriate portions of the general conditions should be deleted in the supplementary conditions and the modified or expanded allowance requirements stated in Division 01.

7.6.2 Specifications

Section 01 21 00—Allowances or the subordinate sections should specify the dollar amount or quantities as well as administrative and procedural matters relating to handling allowances. Specific items covered by each allowance should be identified and may include cost of the product, delivery to site, installation (if applicable), and taxes. Cost items not covered by the allowance should also be identified in the order that they will be included in the contract sum. Specific items not covered by an allowance may include delivery, storage and handling at site, installation of product (if applicable), overhead and profit, and other expenses necessary for a complete installation. These sections should contain appropriate cross-references to sections in Divisions 02 through 49 in which the allowance items are specified.

Sections in Divisions 02 through 49 that are affected by allowances should contain an article in PART 1—GENERAL calling attention to the provisions for allowances. The items covered by an allowance should be cross-referenced to an allowance amount stated in the applicable Division 01 section. In addition, the A/E must provide enough information in PART 2 and PART 3 of each section so that bidders will know exactly what is covered by the allowance. Both the products and their installation should be specified as completely as possible.

Figures 7.6 and 7.8 show how allowances are covered and coordinated in the various contract documents. In this example, both cash and quantity allowances are used.

BID FORM

If quantities for the following allowance items vary from the amounts specified in Section 01 21 29 – Quantity Allowances, the Contract Sum will be adjusted in accordance with the following unit prices:

ITEM DESIGNATION	UNIT OF MEASURE	UNIT PRICE
1. Tenant Corridor Entrance	Each	$ ______
2. Corridor Partition	LF	$ ______

SECTION 01 21 29—QUANTITY ALLOWANCES

1.06 SCHEDULE OF QUANTITY ALLOWANCES

A. Item No. 1 – Tenant Corridor Entrance: Allow a quantity of 55 installed entrances, each to include:
   1. Fire-rated hollow-metal frames, Section 08 11 00 – Metal Doors and Frames:
   2. 1-1/2 hour fire-rated wood doors, 3′-0″ by 7′-0″, Section 08 14 00 – Wood Doors, and Section 08 15 00 – Plastic Doors;
   3. Hardware Set No. 3, Section 08 70 00 – Hardware;
   4. Door sign, Section 10 14 00 – Identification Specialities.

B. Item No. 2 – Corridor Partition: Allow a quantity of 250 linear feet of installed 8′-0″ high partition including:
   1. 3-5/8-inch, 25-gage steel studs spaced 24 inches on center, with two layers of 5/8-inch Type X gypsum board screw attached to each side, Section 09 21 16 – Gypsum Board Assemblies;
   2. Tenant side finish: One coat of primer and two coats eggshell latex paint, Section 09 90 00 – Painting and Coating;
   3. Corridor side finish: Type A vinyl wall covering, Section 09 72 00 – Wall Coverings.

Figure 7.8
Partial Bid Form and Corresponding Division 01 Schedule of Quantity Allowances

7.7 Specifying Alternates

Most master guide specifications and standard documents do not contain extensive provisions for alternates, as they are generally unique to each project. If alternates are to be used, special provisions must be incorporated in the documents to make them effective. Figure 7.9 illustrates how the requirements for an alternate for adding television sets to a hospital project might be stated in the various procurement documents.

Procurement Requirements The solicitation, such as the invitation for bids, should inform bidders or proposers of the request for alternate prices. Solicitation instructions may contain guidance for preparing alternate bids or proposals and state considerations for evaluation and award of contract. Bid or proposal forms should clearly identify alternates with spaces for the respective prices.

Agreement and Conditions of the Contract The executed agreement must indicate, as part of the description of the work, which alternates have been selected by the owner. The contract sum must accurately reflect those decisions. Modifications

PROCUREMENT REQUIREMENTS

Invitation to Bid

Type of Bids: Bids shall be on a stipulated-sum basis of the base contract, and include a separate price for each alternate.

Instructions to Bidders

Each alternate is described in the Specifications and is provided for in the Bid Form.

The price of the Bid for each alternate will be the amount to be added to or deducted from the price of the Base Bid if the Owner selects the alternate.

The Owner may accept alternates in any order; regardless of how they are listed, and determine the lowest responsible bidder on the basis of the sum of the base bid plus any selected alternates.

Bid Form

Alternate No. 1—Patient Room Televisions:
Add ____________________________________ Dollars ($______________)
(In Words) (In Numerals)

CONTRACTING REQUIREMENTS

Agreement

The Contractor shall perform all work required by the Contract Documents for TLC Nursing Home at 123 Primrose Lane, Anywhere, USA, including Alternates No. 1, 2, 3 as described in Section 01 23 00—Alternates, of the Specifications.

SPECIFICATIONS
SECTION 01 23 00

1.06 ALTERNATE NO. 1—PATIENT ROOM TELEVISIONS
A. Provide 72 bedside bracket-mounted television sets on the first floor of "A-Wing" as specified in Section 27 41 19—Portable and Spare Audio Video Equipment.

SECTION 26 61 50

1.04 ALTERNATES
A. Refer to Section 01 23 00—Alternates, for description of work under this Section affected by alternates.

Figure 7.9
Sample Use of Alternates

of standard general conditions are not normally required because alternates listed in the agreement form become part of the work.

Specifications Section 01 23 00—Alternates should identify each alternate and describe the changes of work included in each. Where possible, refer to individual specification sections affected by the alternates. If not shown on the drawings, certain areas or portions of the work may be described to clarify the locations of changes necessitated by the incorporation of alternates. The sections in Divisions 02 through 49 that are affected by the alternates should contain a coordinating or clarifying statement calling attention to the provisions for alternates. Statements should be included in PART 1—GENERAL under the article "Summary."

7.8 Specifying Unit Prices

Whether a project is based on a stipulated sum contract with a few unit prices listed for minor variables or is based entirely on unit prices, the requirements and procedures for handling unit prices must be defined in the procurement documents. These requirements will involve the bid or proposal form along with carefully written Division 01 sections, both of which must be coordinated with the applicable sections in Divisions 02 through 49.

7.8.1 Conditions of the Contract

Specified unit price requirements must be coordinated with statements in the conditions of the contract in order to avoid conflicts, duplication of information, or omission of essential requirements. Most standard general conditions address unit prices. Typically, these requirements need to be expanded in Division 01.

AIA Document A201, and related documents from EJCDC, DBIA, and AGC, include the following paragraphs relating to changes in the work and unit prices:

- AIA Document A201 covers unit price adjustments caused by changes in quantities of magnitude that become a substantial inequity to the owner or contractor and includes methods of determining cost or credit to the owner resulting from a change in the work.
- EJCDC C-700 defines the contract documents as including the contractor's bid when listed and attached as an exhibit to the agreement.
  - This document establishes authority for the engineer to determine actual quantities and classifications of unit price work.
  - This document also establishes the contract price for unit price work as the sum of each estimated quantity multiplied by its unit price.
  - It points out that estimated quantities are not guaranteed but are for the purpose of comparing bids.
  - It stipulates that unit price bids include contractor's overhead and profit. It also covers changes in contract price because of significant differences between estimated and actual quantities to be paid by unit price.
  - This document provides for claims for additional costs.

Review the related DBIA or AGC general conditions to determine the stipulated requirements that are to be a part of the contract language for the project.

If the owner has provided their own documents stating the general conditions that will be used for the project, they will have to be closely reviewed and confirmed with the owner for any related coordination and other project requirements.

7.8.2 Specifications

Section 01 27 00—Unit Prices is generally the place for establishing procedural requirements for unit prices. The section covers methods of determining actual quantities for both progress and final payments. The section must be coordinated with the general conditions, as it may either establish or expand upon the following requirements:

- Unit prices are to include all necessary material, labor, equipment, overhead, profit, and applicable taxes.
- Requirements for the submittal of substantiated measurements of quantities with applications for payment must be defined.
- Quantities set forth in the bid form are estimates based on work shown on the drawings. They are the basis on which bids will be compared and the contract sum determined. Payment will be made for work actually performed and measured. The owner reserves the right to increase or decrease the quantities shown on the drawings by a stated maximum amount.
- The final contract sum will be adjusted by change order to reflect actual quantities for unit price items. Section 01 27 00—Unit Prices must define who is responsible for actual measurement of unit price quantities. If it is the owner or the owner's representative, procedures must be defined for administration and coordination. If the contractor is responsible for measurement, the methods of control and verification by the owner must be specified.

The section may include a listing of unit price items showing the item number, designation, and the unit and method of measurement. Estimated quantities are not required in this section because these will be stated on the bid or proposal form.

Describing the unit price item in this section allows simplification of the bid or proposal form, as only the item designation and number will then be required on the bid or proposal form. Figure 7.10 illustrates a schedule of prices for the bid form and the schedule of unit price items in Division 01 for a civil engineering project. Depending on the project, other Division 01 sections may need to contain requirements or references for unit prices.

If the project contains quantity allowances used in conjunction with unit prices, then either the Level 2 Section 01 21 00—Allowances or the Level 3 Section 01 21 29—Quantity Allowances may contain the following unit price requirement:

> Quantities stated for allowances are estimates for the purpose of equitable bidding and establishing an estimated quantity of work. The owner reserves the right to increase or decrease quantities by a maximum stated amount. The requested unit prices will be used to adjust the contract sum to reflect actual quantities.

Figure 7.8 is a sample partial bid form and the corresponding Division 01 schedule of quantity allowances for tenant corridor entrances and corridor partitions. When alternates are specified, they may be based on stated quantities and unit prices requested. In this instance, Section 01 23 00—Alternates would also need to reference unit prices. It may be necessary to designate which quantities and unit prices are included in alternates.

Matching alternates with applicable quantities may become extremely complicated and must be defined as clearly and accurately as possible. Again, as with the use of allowances, it is advisable to limit the use of alternates.

Specification sections in Divisions 02 through 49 should contain a cross-reference to the applicable section in Division 01 dealing with unit prices. The Division 01 sections should describe how specific unit price items will be measured. These requirements would be in PART 1 under the article "Unit Prices." Care must be used in defining the payment for unit price items. There must be careful correlation with typical details for unit price items shown on the drawings. The method of measurement to be employed must also be clearly stated.

Ordinarily, some items will be measured directly and others will be treated as subsidiary obligations, such as pipe and associated fittings. It is best if the measured quantities

BID FORM

SCHEDULE OF PRICES

ITEM NO.	DESIGNATION	UNIT OF MEASURE	UNIT OF PRICE	ESTIMATED QUANTITY	TOTAL PRICE
	Refer to Section 01 22 00 — Unit Prices		Dollars/Cts		Dollars/Cts
1	Earthwork for Utilities	CY	$_______	563	$_______
2	10-inch Ductile Iron Pipe	LF	$_______	62	$_______
3	30-inch Steel Pipe	LF	$_______	234	$_______
4	30-inch Prestressed Concrete Cylinder Pipe	LF	$_______	115	$_______
5	30-inch Valve and Vault	EA	$_______	2	$_______

TOTAL PRICE:______________________________________ ($_____________________)
(In Words) (In Numerals)

SECTION 01 22 00—UNIT PRICES

1.04 SCHEDULE OF UNIT PRICE ITEMS

A. Item No. 1 Earthwork for Utilities
1. Trench Excavation, select granular backfill, and compaction in accordance with Section 31 23 00—Excavation and Fill.
2. Unit of Measure: Cubic yards of backfill compacted in place.

B. Item No. 2 10-inch Ductile Iron Pipe
1. Pipe and fittings including pressure-testing in accordance with Section 33 10 00—Water Utilities.
2. Unit of Measure: Linear feet of pipe installed.

C. Item No. 3 30-inch Steel Pipe
1. Pipe and fittings including pressure-testing in accordance with Section 33 10 00—Water Utilities.
2. Unit of Measure: Linear feet of pipe installed.

D. Item No. 4 30-inch Prestressed Concrete Cylinder Pipe
1. Pipe and fittings including pressure-testing in accordance with Section 33 10 00—Water Utilities.
2. Unit of Measure: Linear feet of pipe installed.

E. Item No. 5 30-inch Valve and Vault
1. Valve, vault, roadway box, frames, and covers in accordance with DPW Standard No. 890.01 and Section 33 10 00—Water Utilities.
2. Unit of Measure: Each valve with vault installed.

Figure 7.10
Sample Schedule of Prices on the Bid Form and Sample Schedule of Unit Price Items in Division 01

relate directly to the way the work is performed. However, this is not always practical because of the amount of effort required to measure some items of work.

For example, in an earthwork project, the contractor may excavate, stockpile, and eventually place the soil in the finished location. Three unit prices would accurately reflect the work done. However, much more measurement effort is required when three unit prices are used rather than a single unit price for earth fill measured in place. The extra measurement effort for three unit prices reduces the risk that the contractor may be overpaid, but the owner must pay for the additional measurement effort. Therefore, earthwork is usually measured in place, either in the finished work or from the borrow pit. Truckload and loose volume measurements should be avoided whenever possible.

Chapter 8
Coordinating Drawings and Specifications

8.1 Introduction to Coordinating Drawings and Specifications

The drawings and the specifications are complementary and both are needed to fully describe a construction project. The drawings show size, form, quantity, relationship, generic type, and graphic representation of construction materials. The specifications define the qualitative requirements for products, materials, and workmanship upon which the construction contract is based. The specifications also describe administrative procedures regarding both drawings and specifications. The drawings and specifications may separately describe limits of work or construction phases.

Aspects of the specifications and the drawings may be clear, concise, correct, and complete as to the information they convey, but unless the parts are coordinated with each other, the construction project may experience many problems and discrepancies. Effective coordination and quality assurance programs must begin early in the design process and depend largely on continuous and effective communication among project team members.

Coordination must occur at all levels of the project team. Consultants must ensure that the drawings and specifications within their offices are fully coordinated. Various consultants must also coordinate their graphic and written materials among their respective offices. The responsible A/E must not only coordinate the drawings and specifications within the A/E's own office, but also be responsible for coordinating the entire construction document package and for maintaining the communication process that will facilitate this coordination.

8.1.1 Drawings

A particular material or component may appear many times throughout the drawings, but is specified in only one location. In order to simplify and coordinate the documentation process, only generic notes should be used on the drawings to identify, but not describe, a material or component. Overly detailed notes may obscure the drawings and increase the possibility of inconsistencies and duplications.

Detailed written information should be reserved for the specifications so that minor changes during development of the documents can be accommodated by revising only the specifications. For example, if the drawings have been noted to include concrete pavers, a design decision changing to brick pavers would necessitate finding and changing several drawing notes as well as the specifications. However, if the drawings had indicated only unit pavers, those notes would be unaffected by the change, and only the specifications would have to be modified.

Minimizing the number of required changes to written information on the drawings diminishes the opportunities for discrepancies among the various drawings and between the drawings and specifications. Whenever a change of materials is made, the corresponding symbols must be changed throughout the drawings, so symbols referring to a generic class of material, such as concrete or wood, are preferable over symbols referring to specific materials, such as walnut and oak. Thus, a change in wood species, for example, would not affect the drawings. Also, when more than one type of a generic class of materials is used, then such differentiation can be shown in the drawings with short notation such as "Wood-A" and "Wood-B." Items of equipment, such as pumps or valves, should be identified on the drawings by a short generic name or coded symbol. For example, power roof ventilators (PRVs) might be indicated by PRV-1 and PRV-2. The graphic representation on the drawings should be only a representative outline rather than a detailed drawing.

Proprietary product information should not appear on drawings unless a closed proprietary specification limits options to a single product. Even when specifications refer to products by their proprietary names, the notes on drawings should remain generic. If proprietary names are used on drawings, acceptance of a substitution could require revision of each proprietary reference by addendum or contract modification.

Material systems should also be identified on the drawings with only generic notes. For example, either a four-ply built-up roof with gravel surfacing or a loose-laid and ballasted ethylene, propylene, diene monomer (EPDM) roof could be noted on the drawings simply as "roof membrane." If more than one type of membrane is used in a project, the drawings should identify roofing Type A, B, C, and so forth, where the letter symbolizes a specific system described in the specifications.

Where different design strengths are specified for structural steel, reinforcement, concrete, or other structural elements, the drawings may be used to indicate the boundaries between the materials of different capacities, which also may be designated as Type A, Type B, and so on.

Do not cross-reference between drawing and specifications with notes such as "refer to specifications" or "manhole cover—see specifications." Drawings and specifications are complementary parts of the same set of contract documents, and including references from one to the other is not recommended. However, it is acceptable to refer to specific specification sections. For example, a note on a floor plan could state "provide firestopping at wall penetration—see specification Section 07 84 00." This concept is consistent with similar techniques on drawings, where one drawing references another specific drawing or detail in the drawing set.

The drawings should not attempt to define the work of specific subcontractors or trades. However, the drawings may be used to indicate the extent of alternates, areas of construction phasing, limits of work, and specific items of work by the owner or by separate contract.

On multiple-prime contract projects, the drawings can be used to designate work of separate contracts. Although these items may be graphically delineated on the drawings, the written descriptions and scope of work requirements should be defined in Division 01—General Requirements and PART 1—GENERAL of the specification section.

8.1.2 Specifications

In contrast to the generic notes and symbols on the drawings, the specifications provide detailed requirements for the physical properties, chemical constituents, performance requirements, and standards of workmanship associated with the manufacture and installation of materials, equipment, and components. For example, whereas the drawings might simply refer to a vapor retarder, the specifications define minimum perm rating, physical properties, and required installation methods.

The specifications usually do not, however, include information that belongs more appropriately on the drawings, such as quantity, capacity, and location.

8.1.3 Schedules

Schedules help simplify communication by presenting data in a tabular form or in a matrix. The location and content of schedules may vary widely among architects/engineers (A/Es). When placed in a specification section, schedules are included at the end of PART 3—EXECUTION. Although not technically part of execution, the schedules are placed there for convenience in specification preparation, although they can be included in the specifications or on the drawings.

Examples of schedules are hardware schedules and sealant schedules. Schedules that include materials from multiple specification sections should be included in the drawings.

8.1.4 Coordination

Specifications complement, but should not repeat, information shown on the drawings, nor should the drawings duplicate information contained in the specifications. If a requirement on the drawings or in the specifications is duplicated in the other, an opportunity arises for a discrepancy between the two. An addendum covering a design change may correct the item in one location but overlook it in the other. Last-minute changes are most likely to create discrepancies of this sort. Such discrepancies may cause bidders/proposers to make different interpretations of what is required, often resulting in change orders and extra costs. Properly prepared drawings and specifications should dovetail like a jigsaw puzzle, without overlaps or gaps.

To facilitate coordination, especially on complex projects with many different materials, components, subcontractors, and trades, adhere to the CSI principles established for locating information properly within the contract documents.

8.1.4.1 Preliminary Coordination

Construction documents are generally reviewed at set milestones such as 50 percent completion and 90 percent completion. Drafts of interim specification sections and copies of interim drawings from consultants should be collected and reviewed for coordination of information:

- Within the project manual
- Within the drawings of each separate discipline
- Between the drawings of separate disciplines
- Within the specifications of each separate discipline
- Between the specifications of separate disciplines
- Between the project manual and drawings

Consultants should review the drawings and specifications of other disciplines and forward corrections and comments to the A/E for communication to the other project team members.

In addition to obvious graphical, dimensional, and typographical errors, documents should be carefully checked and compared to eliminate:

- Omissions
- Overlaps and duplications between disciplines
- Noncompliance with laws and regulations
- Conflicts and discrepancies with locations of equipment and components
- Incompatible materials and components
- Difficult or impossible construction methods
- Inconsistent terminology and abbreviations
- Inconsistent units of measure
- Incorrect or unspecified materials, components, or equipment
- Errors in extent of alternates
- Errors in defining areas of construction phasing
- Errors in defining limits of work
- Errors in identifying work by the owner or work not in contract
- Errors in designating work of separate contracts
- Inaccurate or unnecessary cross-referencing

Checking construction documents is not the same as coordinating the separate portions. A drawing may be found correct unto itself but may be totally inconsistent with other parts of the construction documents. Particular attention should be given to coordinating Division 01 sections with the conditions of the contract. The level of specification detail should complement the level of drawing detail. Early coordination with sufficient time devoted to the task will help achieve design intent and promote elimination of problems during the construction stage.

8.1.4.2 Final Coordination

Final construction documents should receive final coordinating and checking prior to issuance:

- Verify that all previously noted inconsistencies, errors, and inaccuracies have been corrected.
- Verify that the construction documents are complete, with final check of project manual table of contents and drawing table of contents.
- Verify consistency of all schedules.

8.2 Terminology and Abbreviations

Use consistent terminology and abbreviations throughout the contract documents to avoid confusion among the various users. The "service sink" in the specifications should not become the "janitor's sink" on the drawings, nor should "bituminous surface course" become "asphalt topping."

The use of inconsistent terminology and abbreviations may create confusion in preparing specifications. It wastes time and can easily raise construction costs and cause delays, and may even result in faulty construction.

The A/E should take responsibility for enforcing consistent terminology and abbreviations. Inconsistent and inappropriate terminology is perpetuated by such poor practices as the indiscriminate reuse of detail drawings and poorly edited specifications from previous projects.

8.3 Precedence

Questions of precedence often arise among large-scale drawings, small-scale drawings, and specifications. However, the use of statements in the project manual to establish the precedence of the various contract documents is not recommended because the documents are complementary. The general conditions of the contract should indicate only that in a case of conflict between drawings and specifications, the A/E will make a documented interpretation.

8.4 Responsibility for Coordination

One person should be the coordinator and should have the responsibility for reviewing consistency between the drawings and the specifications. Good coordination policies include establishing effective intra-office and inter-team communication procedures. In preparing drawings and specifications, the specifier and A/E should work together, keeping in mind the difficulties faced by estimators, bidders, contractors, inspectors, and product representatives.

In many instances, the A/E is also the specifier. As the project design stage begins, a conference should be held that includes the A/E, the consultants, the drafters, and the specifier. A preliminary project description or outline specification should be developed as a checklist, and additional conferences should be held as the design progresses.

Specification preparation should proceed concurrently with design, and coordination should become a continuous process from the beginning of the project.

8.5 Using Checklists

The coordinator should use checklists to ensure that necessary items are included in the specifications, that specified items are consistent with the drawings, and that drawings do not duplicate the specifications.

The form of a checklist will vary depending on its purpose.

Forms may be designed to address specific office problems such as omissions, discrepancies, or duplications. For example, rather than putting overly specific or detailed notes on a drawing, drafters and computer-aided design (CAD) operators might simply keep a list of catalog references used in preparing details of particular products and equipment. These lists may be maintained in specification outlines that the specifier can later consult when preparing each section.

A coordination checklist designed to avoid omissions in the specification of essential items noted in the drawings might have lists of materials, systems, and products by specification system. Check marks can be used to indicate inclusion in the design, and space should be provided for notes regarding coordination with other work and other items requiring special attention. *MasterFormat®* is one basis for an effective coordination checklist. Figure 1.3 (Chapter 1) shows a portion of such a checklist for items in Division 08. The designer or drafter should fill in appropriate information on the list as work progresses to inform the specifier of required items.

Chapter 9
Modifications

9.1 Procurement and Contract Document Modifications

There is no effective substitute for clear, concise, correct, complete, and carefully coordinated construction documents. Despite best efforts and planning, situations requiring changes do develop. The architect/engineer (A/E), owner, and contractor must have adequate means for dealing with such changes. Addenda and change orders may be employed to modify procurement and construction documents, but reliance on them to correct poorly prepared drawings and project manuals should be avoided. Addenda are modifications issued either prior to receipt of bids and proposals or, in some cases, prior to the time the agreement is executed. Change orders are modifications issued after the agreement is executed.

Although many public agencies have established forms and procedures for addenda and change orders that may differ slightly from suggestions contained in this chapter, the differences are procedural and do not affect stated principles.

9.2 General Criteria

Written construction document modifications are the means for communicating changes to the bidders or proposers during procurement and to the contractor during construction. Most changes are best expressed in words and augmented by drawings where necessary. However, it is sometimes difficult, if not impossible, to change complex drawings by written statements alone. Oral instructions or changes should be avoided, even where it is intended that an addendum or change order will be prepared later to cover the change. Instead of giving an oral interpretation, an A/E should state that an addendum or change order will be issued promptly to clarify the items in question.

Most instructions to bidders or proposers state that oral interpretations are not binding, and the general conditions usually require all contract modifications to be in writing.

There are two basic methods for preparing written changes to procurement and construction documents: the narrative method and the revised page or drawing method.

9.2.1 Narrative Method

In the narrative method, the following criteria should be observed:

- *Brevity.* Give only enough information to make the change clear, but avoid oversimplification.

- *Clarity.* Give enough information to clearly communicate the required changes. When modifying written documents, repeat enough of the original text to make each change self-explanatory. In cases where only a few words are added to or deleted from long sentences, or where changes become voluminous, or where it takes much longer to explain the change than to actually make the change, cross-referencing to original documents without repeating the original text may be preferable. If necessary, new sections can be added to the specifications. When modifying graphic documents, give the location of the change by using the drawing number, drawing title, and location on the drawing, such as plan title, detail title, schedule, or column line coordinates. In any event, one principle should govern the method of preparation: emphasize the changes.
- *Cross-References.* Express a change only once, and then refer to it wherever necessary. Where a change in specifications requires a change in drawings, explain all related changes in the addendum or change order. Thereafter, refer the reader to the addendum or change order rather than repeating the instructions. One correction properly expressed should serve for all repetitive changes. Verify that changes do not contradict or make some other part of the work impossible to accomplish.

9.2.2 Revised Page Method

Depending on the number and type of changes, the revised page method may be a more effective means of communicating the changes to the project manual or the drawings. This method of preparing changes to written and graphic documents is a reissuance of affected pages or sheets with appropriate markings to identify the changed portions. Although this method may not save time in preparing the change, it does save time in posting it.

One procedure for handling page revisions to the project manual is placing a mark or other symbol in the margins on either side of the original word(s), line(s), or paragraph(s) in the previous issue and indicating the number of the addendum or modification that describes the change.

Another procedure entails using a different type of printer font, italicizing, or shading to highlight the revised words. Deleted words can be lined through or crossed out. Revised pages are referenced in the addendum or modification and should be listed as well as assembled in the order in which they appear in the project manual. Each of the revised pages should include a revision date. This method minimizes time spent posting changes and reduces the possibility of inaccurate or incomplete posting.

Drawing revisions can be made either by issuing a sketch on a sheet size smaller than the original drawing or by revising a portion of the drawing and reissuing the entire drawing. The change should be clearly identified. In the case of a sketch, it should refer to the original drawing and should include a revision date. In the case of a revised drawing, the change should be visually identified and a revision number placed adjacent to it. The revision number should be coordinated with a short explanation in the appropriate place on the drawing title block along with a revision date.

In either case, a list of sketches or a list of revised drawings should be attached to the construction document modification. The method of reissuing entire drawings, although more expensive from a printing standpoint, allows substitution of complete drawings into sets and eliminates the disadvantage of working with both drawings and sketches.

9.3 Procurement Documents Modifications

9.3.1 Addenda

Addenda are written or graphic instruments issued to clarify, revise, add to, or delete information in the procurement documents or in previous addenda. Typically, addenda are issued before the opening of bids/proposals.

The American Institute of Architects (AIA) and Engineers Joint Contract Documents Committee (EJCDC) general conditions differ slightly in their approach to addenda. *AIA Document A201, General Conditions*, states that addenda are instruments issued before execution of the agreement. This allows for negotiated adjustments of a bid/proposal by issuance of an addendum. In contrast, *EJCDC C-700 General Conditions* define addenda as instruments issued prior to receipt of bids/proposals.

The AIA Document A201 definition reflects the perspective of the private-sector owner who is not bound by public policy and can negotiate and issue addenda right up to the time of execution of the agreement. However, the primary purpose and use of an addendum is to make changes before opening of bids/proposals.

The EJCDC C-700 definition reflects the view of the public works projects that may preclude modification of procurement documents between bid opening and execution of the agreement. Some jurisdictions allow negotiation with the lowest responsible bidder. Refer to Figure 9.1 for a graphic illustration of this concept.

9.3.2 Purpose

The primary purpose of addenda is to clarify questions raised by bidders or proposers; to issue new requirements, including changes to the extent of work; and to correct errors or omissions in procurement documents. Addenda are used when it is necessary to change the documents in any way. Examples include the following:

- To change date, time, or location of receipt of bids or proposals
- To add to, delete from, or revise procurement documents
- To include additional qualified products or approved substitutions

References to addenda, the method of issue, and other pertinent facts concerning addenda are included in the instructions for procurement, the bid/proposal form, the agreement, and the conditions of the contract.

When issued, addenda become part of the procurement documents, and those portions of addenda that affect the contract documents become part of the contract documents after the agreement is executed.

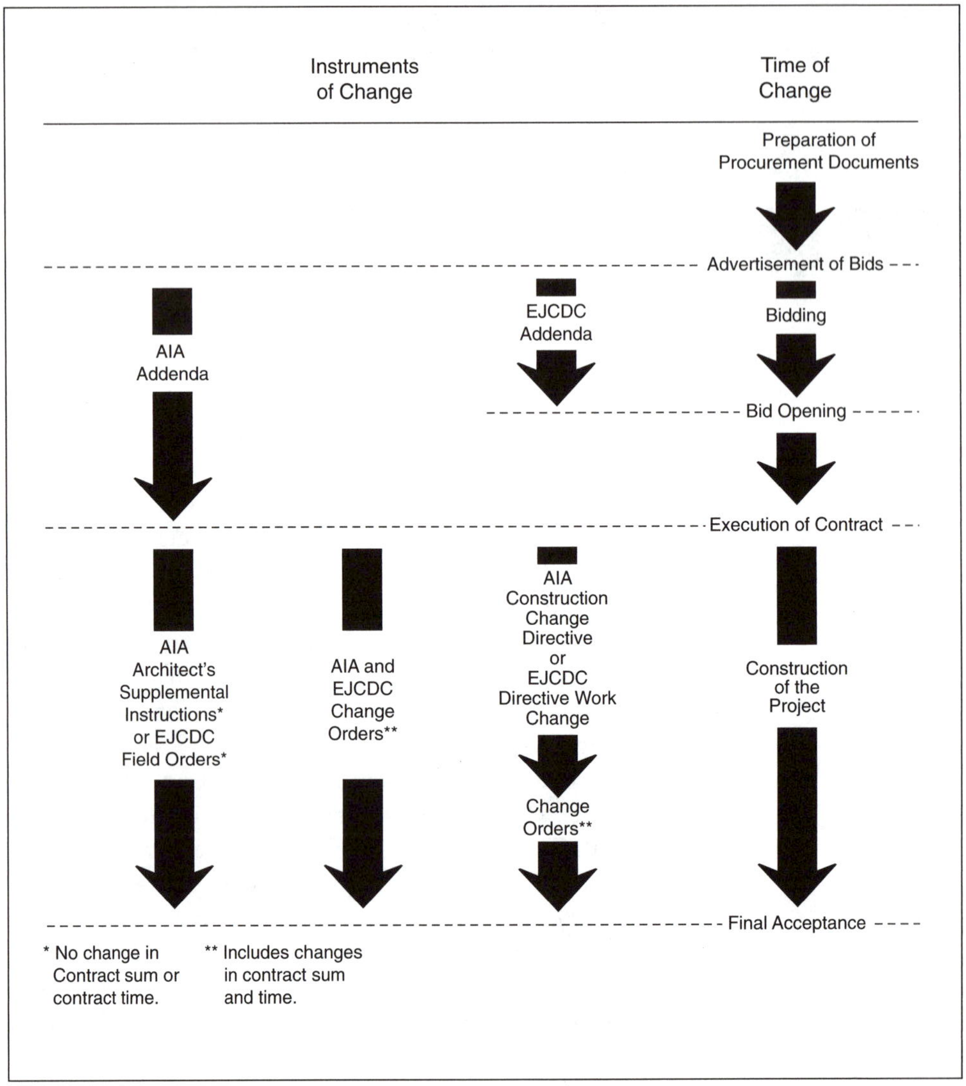

Figure 9.1

Changes in Procurement and Contract Documents

9.3.3 Procedures

The value of the work should determine whether a change should be made. It may sometimes be advantageous not to issue addenda for noncritical changes that do not affect cost or time, and defer them until after the agreement is executed. Numerous minor changes can be covered with simple statements.

Addenda should be numbered consecutively. Where separate contracts are to be awarded, a separate series of addenda numbers for each contract should be used. Where feasible, a simple numbering system for items within an addendum helps future cross-referencing.

In the procurement process, addenda should be issued to bidders/proposals in time for use in preparing prices. If an unforeseen but critical question arises, it may be wise to issue an addendum, even if it means delaying the bid/proposal opening.

A minor addendum may sometimes be issued as late as four days before the date on which bids are due without imposing a hardship on the bidders/proposals.

However, standard instructions to bidders usually establish time limits within which the bidder/proposer may submit questions or request substitutions.

- *AIA Document A701, Instructions to Bidders*, stipulates at least 7 days prior to date of receipt of bids.
- *EJCDC C-200, Guide to Preparation of Instructions to Bidders*, recommends that questions should be submitted at least 10 days prior to opening of bids.

Attempts to make major last-minute changes by issuing a few simple inadequate statements in a hastily prepared addendum should be avoided. If it is imperative that a last-minute addendum describing a major change be issued, the owner should be advised and, if possible, the bid/proposal opening date should be postponed. Addenda issued in a proposal process should follow the same guidelines.

Bidders and proposers often need to forward copies of addenda to sub-bidders; therefore, addenda should be easily reproducible or in electronic form. If addenda are to be issued electronically, they should be produced in an electronic format that prevents modification, such as a portable document format (PDF).

9.3.4 Addenda Format

Written addenda may be accompanied by drawings. Addenda items should be arranged in the same sequence as the project manual and drawings. Basic components of addenda and the order in which they should appear are as follows:

A. Addenda number and date of addendum
B. Project identification
C. Name and address of A/E
D. To whom addendum is issued
E. Opening remarks and notice to bidders or proposers
F. Changes to prior addenda
G. Changes to procurement requirements
H. Changes to contracting requirements
I. Changes to specifications, in sequence, including approved substitutions
J. Changes to drawings, in sequence

9.3.5 Sample Addendum

The letters shown on the left side of Figure 9.2 corresponds to the 10 basic components of an addendum format, as listed above. This sample has been limited to one page, although an actual addendum may require many pages. Each item should be numbered consecutively for ease of reference.

A. Number and Date →	**ADDENDUM NO. 2, May, 3, 2010**
B. Project identification →	RE: First National Bank of Brownsville project No. 11863
C. Name and Address of A/E →	FROM: JONES and SMITH, Architects and Engineers John Doe Building Washington, D.C. (202) 555-8888
D. To Whom Addendum Is Issued →	TO: Prospective Bidders
E. Opening Remark and Notice to Bidders →	This Addendum forms a part of the Contract Documents and modifies the original Procurement Documents dated April 15, 2010 and Addendum No. 1. dated April 20, 2010, as noted below. Acknowledge receipt of this Addendum in the space provided on the Bid Form. Failure to do so may subject Bidder to disqualification. This Addendum consists of one page and the attached drawing, Sheet S-9-A, with the revised date of April 20, 2010.
F. Changes to Prior Addenda →	CHANGES TO PRIOR ADDENDA: 1. Addendum No. 1, Item No. 13, Page AD-1-I. change the number of the referenced drawing from "G-I" to "G-7".
G. Changes to Bidding Requirements →	CHANGES TO PROCUREMENT REQUIREMENTS: 2. Instructions to Bidders, Page 00 21 13-2, Item 15, replace first sentence with "Proposed substitutions must be submitted in writing at least 15 days before the date for opening of bids"
H. Changes to Contracting Requirements →	CHANGES TO CONTRACTING REQUIREMENTS: 3. Supplementary Conditions Item No. 12, Page 00 73 00-3, change limit of public liability from "$100,000/$500,000" to "$300.000/$500.000."
I. Changes to Specifications →	CHANGES TO SPECIFICATIONS: 4. Section 09 21 16–Gypsum Board Assemblies, subparagraph 2.01 C.3, add the following: "3. Smooth shank nail: a. ASTM C 514 b. Length: 1-3/8 inches (35mm)" 5. Section 23 61 00—Refrigeration Equipment subparagraph 2.03 B. 2, change total square feet (square meters) of surface from "298 (27.68)" to "316 (29.36)."
J. Changes to Drawings →	CHANGES TO DRAWING: 6. Delete sheet S-9—Beam Schedule and replace with attached Sheet S 9–A. 7. Sheet M-1—Mechanical Plan. at Room 602, change "12 × 6" exhaust duct, to "12 × 18"
	END OF ADDENDUM

Note: This sample is reasonably complete. It touches on all major points included in most Addenda. The user should recognize that such project may have specific requirements that may not be emphasized in this sample document.

Figure 9.2
Sample Addendum

9.4 Contract Documents Modifications

The respective EJCDC and AIA conditions of the contract specifically set forth procedures for the A/E to issue interpretations of contract documents and to order minor modifications of the work consistent with these interpretations. Provisions and procedures are also included for the contractor and owner to claim changes in contract sum or contract time if the contractor or owner believes such changes are justifiable based on the A/E's written interpretation.

9.4.1 Purpose

Frequent reasons for issuing modifications include:

- Incorporation of changes in owner's requirements, including additions or deletions in the work
- Unforeseen field conditions that necessitate changes in the work
- Changes in regulatory code provisions, or other requirements of authorities having jurisdiction, which require changes in work
- Changes in market conditions (e.g., specified products are not available, new products are considered that offer price advantages or other benefits, or new information is made available that affects specified products)
- Correction of contract documents to eliminate errors, omissions, or discrepancies

9.4.2 Procedures

To minimize delays and costs of changes, contract document modifications should be issued in a timely and orderly fashion. A change initiated early in the construction process allows time to consider proposals and alternatives and to determine a mutually acceptable change without affecting work in progress.

Contract document modifications can usually be kept concise by referring to attachments that describe changes and cost breakdowns in detail.

Procedural requirements for initiating and processing modifications should be specified in Division 01, Section 01 26 00—Contract Modification Procedures, and closely coordinated with the conditions of the contract and the agreement.

The method used to describe contract document modifications may be similar to the method used for an addendum.

9.5 Methods of Modification—AIA Documents

There are three recognized ways to modify contract documents when using AIA Document A201:

- Change order
- Construction change directive
- Minor changes in the work (Architect's Supplemental Instructions)

9.5.1 Change Order

A change order is a written instruction to the contractor issued after execution of the agreement that authorizes an addition, deletion, or revision in the work or an adjustment in the contract sum or the contract time. Change orders must be signed by the owner, the A/E, and the contractor. The contract sum or time can be adjusted only by written change order. The standard form for change order is *AIA Document G701, Change Order.*

According to AIA Document A201, a change order is a written instrument prepared by the architect and signed by the owner, contractor, and architect, stating their agreement on all of the following:

- A change in the work
- Amount of the adjustment, if any, in the contract sum
- Extent of the adjustment, if any, in the contract time

9.5.2 Construction Change Directive

A construction change directive is a means of ordering changes in the work when the owner and contractor have not agreed on the proposed changes in contract sum or time. The standard form is *AIA Document G714, Construction Change Directive.*

According to AIA Document A201, a construction change directive is a written order prepared by the architect and signed by the owner and architect, directing a change in the work prior to agreement on adjustment and stating a proposed basis for adjustment, if any, in the contract sum or contract time, or both. With a construction change directive, the owner may, without invalidating the contract, order changes in the work within the general scope of the contract consisting of additions, deletions, or other revisions, the contract sum or contract time being adjusted accordingly.

A construction change directive is used in the absence of total agreement on the terms of a change order. Upon receipt of a construction change directive, the contractor is required to perform the changes and advise the owner and architect of agreement or disagreement with the proposed method for adjusting contract sum or time.

If the contractor agrees with the proposed method for adjusting the contract sum or time by signing the construction change directive, the agreement should be recorded as a change order.

If the contractor disagrees with the proposed method for adjusting the contract sum or time, the work must still be performed and the architect is usually required to determine the method and amount of adjustment on the basis of reasonable expenditures and allowance for overhead and profit.

9.5.3 Minor Changes in the Work

Minor changes in the work are defined as instructions or interpretations not involving adjustment to the contract sum or time. Such changes are made by the architect's supplemental instructions directing the contractor to make stated modifications. The architect's supplemental instruction does not require the owner's signature. The standard form is *AIA Document G710, Architect's Supplemental Instructions.*

Under A201 the architect will have the authority to order minor changes in the work not involving adjustment in the contract sum or extension of the contract time and not inconsistent with the intent of the contract documents. Such changes shall be effected by written order and shall be binding on the owner and contractor. The contractor is required to carry out such written orders promptly.

For A/Es working on projects with construction management, refer to *AIA Document G701/CMa, Change Order—Construction Manager–Adviser Edition*, for use with *AIA Document A201/CMa, General Conditions of the Contract for Construction—Construction Manager–Adviser Edition*.

9.6 Methods of Modification—EJCDC Documents

There are four recognized ways to change contract documents when using EJCDC C-700:

- Change order
- Work change directive
- Field order
- Written interpretation or clarification

9.6.1 Change Order

Change orders must be signed by the A/E recommending the change, the owner authorizing the change, and the contractor accepting the change. The contract price or time can be adjusted only by written change order. The standard form for change order is *EJCDC C-941, Change Order*.

EJCDC C-700 defines a change order as a document recommended by the engineer, signed by the contractor and the owner, and authorizing an addition, deletion, or revision in the work or an adjustment in the contract price or the contract time, issued on or after the effective date of the agreement.

9.6.2 Work Change Directive

The work change directive may or may not affect the contract price or time but is evidence that the changes will be incorporated in a change order once the value of the work is established. The standard form is *EJCDC C-940, Work Change Directive*.

The EJCDC C-700 defines a work change directive as a written directive statement to the contractor issued on or after the effective date of the agreement and signed by the owner and recommended by the engineer ordering an addition, deletion, or revision in the work, or responding to a differing or unforeseen subsurface or physical condition under which the work is to be performed or to emergencies.

A work change directive will not change the contract price or the contract times, but is evidence that the parties expect that the change directed, ordered, or documented by a work change directive will be incorporated in a subsequently issued change order following negotiations by the parties as to its effect, if any, on the contract price or contract times.

9.6.3 Field Order

EJCDC C-700 allows the engineer to authorize or order minor variations in the work when such changes do not involve a change in contract price or contract time. These orders must be in writing, but do not require an executed change order or the owner's signature. EJCDC does not have a standard form for field orders.

Under EJCDC C-700 a field order is a written order issued by the engineer that orders minor changes in the work but that does not involve a change in the contract price or contract times.

9.6.4 Written Interpretation or Clarification

EJCDC C-700 defines a written amendment as a written interpretation or clarification of statement modifying the contract documents, signed by the owner and the contractor on or after the effective date of the agreement, and normally dealing with the nonengineering or nontechnical rather than strictly construction-related aspects of the contract documents.

9.7 Methods of Modification—DBIA Documents

DBIA Document No. 535—Standard Form of General Conditions of a Contract Between Owner and Design-Builder recognizes three ways to change the contract documents:

- Change order
- Work change directive
- Minor changes by written notice

9.7.1 Change Order

Change orders are signed by the owner and design-builder to document changes in the scope of the work, amount of adjustments to the contract price, and extent of adjustment to contract time.

9.7.2 Work Change Directive

A written order prepared and signed by the owner, directing a change in the work prior to agreement of an adjustment in the contract price or contract time. The general conditions require both parties to negotiate in good faith on reaching agreement. Final agreement on cost and time is documented in a written change order.

9.7.3 Minor Changes by Written Notice

DBIA general conditions allow the design-builder to make minor changes in the work as long as the changes do not materially or adversely affect the design, quality, performance, and workmanship required by the contract documents. Changes must be consistent with the intent of the contract documents, and the design-builder must inform the owner promptly in writing.

9.8 Methods of Modification—Other

There are other entities that produce general condition contract documents such as Associated General Contractors (AGC), the General Services Administration (GSA), the U.S. Department of Defense (DOD), state and other public works, and so on. If, at the direction of the owner-client, a standard construction document is not being used, care must be given to confirm the method and procedures along with supporting documents and forms to manage any modifications that may be required during the performance of the construction phase of the contract.

Pay close attention to any special requirements that may also apply to any modifications to the documents prior to final bidding or procurement negotiation of the contract for these nonstandard contract documents.

9.9 Change Orders

An axiom of both AIA Document A201 and EJCDC C-700 is that only the owner has authority to execute a change order.

Execute may be defined as authorizing the change order.

The A/E signs the change order recommending the change and the contractor signs the change order indicating acceptance of the change. Change orders may originate either by issuance of an AIA construction change directive or an EJCDC work change directive or by a change order proposal request.

A change order proposal request can be initiated by the A/E by issuing an *AIA Document G709, Work Changes Proposal Request.* A change order proposal request describes proposed changes, supplemented, if required, by additional or revised drawings and specifications. It solicits a proposal from the contractor for adjustment (if any) of the contract sum or contract time or both. A proposal request is for information only and is not an instruction to execute the proposed changes or to stop the work in progress. The contractor will then respond to the owner regarding adjustment of the contract sum, contract time, or both. Upon agreement of terms, the A/E writes the change order, which is signed by the owner, cosigned by the A/E, and then signed by the contractor indicating acceptance of the agreed-upon change.

Chapter 10
Master Guide Specifications

10.1 Introduction to Master Guide Specifications

Most architect/engineer (A/E) firms, government agencies, and large corporations use master guide specifications as a basis for their individual construction documents. Developing and using master guide specifications is an efficient way of producing project specifications.

For any given specification section, there are two basic methods of preparation of text: (1) write the text from scratch or (2) edit prewritten text. Obviously, option two is usually less time consuming, but it has its drawbacks, too. When using prewritten text, A/Es have a number of sources to draw from, including commercial master guide specification services, office master text, specifications furnished by product manufacturers, and previous project specifications. Each text source has its pros and cons, but by far the most effective and efficient source is a set of text that is pre-edited to the A/E's principal project types and specification requirements.

The term *master specifications* refers to the documents used as guides for preparing project specifications. The U.S. Department of Defense (DOD) and Department of Housing and Urban Development (HUD) refer to their specifications as guide specifications. These manuscripts facilitate the preparation of project specifications by standardizing products, materials, and processes and their order of presentation. These specifications allow editing to adapt the guide specification to specific project requirements. These two terms are combined to become the term *master guide specifications,* used in this manual for prepared specification sections intended to be edited and used as project specifications.

The term *office master specifications* is also used, principally to distinguish an A/E's master specification from commercial master guide specifications. The office master specification may be derivative of commercial master guide specifications that were used as its basis.

A master guide specification ideally would include the types of items typically utilized for most projects. Each master guide specification section should include text written in a consistent style covering typical requirements and should also list possible options and choices. Instructional notes may be included and should provide direction, guidance, and notice of required decisions. The A/E edits selected master guide specification sections to suit the particular project. Paragraphs and articles that do not apply are deleted.

Master guide specifications have evolved from the need to compile a concise, centralized, and quick-to-edit source of information from the overwhelming amount of data and options available to the A/E. When new project experience is systematically incorporated into the master for future use, the master guide specification becomes a significant repository of knowledge for the firm (i.e., a part of the "corporate memory"). A master guide specification prepared and maintained with an emphasis on consistency of text and speed of editing makes project specifications more accurate and quicker to prepare. Investments in the firm's master guide specification will be returned through lower project specification preparation costs.

Using a master guide specification does not eliminate the need to understand the principles of specification writing. Master guide specifications are only tools. They require competence, skill, and construction experience for their proper use. Although master guide specifications are extensive, they might not cover every need of a specific project. These additional needs will require custom specifications to be written consistent with the master guide specifications. The use of a master guide specification requires an understanding of the basis of the conditions of the contract and general requirements for which they were prepared.

10.2 Contents of Master Guide Specifications

A complete master guide specification system should consist of carefully coordinated elements that together form a total information library, including the following:

- Master guide specification sections themselves
- Provisions for coordination of drawings and specifications
- List of frequently specified products and additional required information
- Checklist for each specification section, identifying principal decisions needed for the individual project
- Information and evaluation about products, materials, systems, codes, and standards, utilized in the master guide specification

10.3 Reasons for Using a Master Guide Specification

The reasons for developing and using a master guide specification or purchasing a commercially available system for a firm's project specifications include the following:

Easier Updating and Maintenance of Specification Data A master guide specification is an efficient method for incorporating and recording changes in product data and administrative procedures as needed to keep specifications current.

Improved Efficiency in Specifying Project time can be saved by using a master guide specification. Repetitive development of the same specifications for multiple projects is virtually eliminated. Although overhead costs for developing (or purchasing) and maintaining the office master guide specification must be charged against project savings, a net saving can easily be realized. Timesaving tools, whether purchased or developed in-house, are feasible only with carefully coordinated, consistently formatted documents.

Expanded Decision-Making Capability A master guide specification that presents a wide range of standard options enables the A/E to make the best possible choice for each item in a project.

Avoidance of Delays in Project Development Use of a master guide specification allows specification production concurrent with drawings, avoiding the delay, confusion, and misuse caused by last-minute preparation and production.

Minimized Repetitive Work A master guide specification can free personnel from the effort of developing a specification from scratch for each project, allowing them to concentrate on more important technical requirements.

Reduced Errors and Omissions Inadvertent mistakes and omissions can be minimized through use of a carefully developed and maintained master guide specification system. A/Es become familiar with the location of specific topics in the specification. New projects are started from the definitive databank that records feedback from projects and updates received from other sources, such as from a commercial subscription service. By reducing the time required to prepare project specifications, personnel can concentrate on properly addressing critical issues. Electronic automation tools further reduce errors and omissions by automatically performing programmed operations. The office master specification provides a vehicle for capturing experience and developing a corporate knowledge base.

Reduced Exposure to Liability By reducing errors and omissions, a master guide specification may minimize the potential for litigation.

Standardized Office Policies and Procedures A master guide specification system can be instrumental in standardizing technical terminology, scope of service to clients, and office policy on a wide range of technical considerations. Office unified technology should be based on Construction Specifications Institute (CSI) formatted documents including *CSI/CSC MasterFormat®, UniFormat™,* U.S. National CAD Standard for Architecture, Engineering, and Construction (NCS), and other developing integration/standardization industry efforts to organize data.

Improved Office Practices A master guide specification can be used to train inexperienced personnel and establish a standard production procedure. Expanding and updating the master during slow periods can level the workload and initiate systematic office improvements. Many firms use the master guide specifications in their in-house educational programs. Using personnel during slow periods to update and maintain the master can spur improvements as well as educate.

Electronic Technology to Enhance Production and Improve Efficiency Master guide specifications can enhance production by incorporating internal software linkages that provide coordination and assurances in choices that prevent contradictory information. Department of Defense Unified Facilities Guide Specifications and some commercial specification systems provide options for generating tables of contents, listings of submittals, reference standards, and other reports. Electronic software permits efficiencies with macros and hyperlink features that can provide additional benefit beyond the printed text. Electronic technology can provide tools for coordination of specifications and drawings and help reduce the potential for errors and omissions. Additional considerations for electronic technology are included at the end of this chapter.

Building Information Modeling (BIM) Systems and programs are under development to incorporate building models and objects into three-dimensional representations of the project. These systems will assist with the coordination of all the assemblies and components as they are designed by the engineers and consultant. The supporting documentation and specifications will be included in the data file system being developed for the project. These programs are in the early development stages, and as they are released the specifications will require additional document management to be incorporated in the project.

The same specifying techniques described for project specifications apply to master guide specification preparation with the addition of considerations relating to:

- Use for more than one project type, client or client type, location, or code jurisdiction
- Alternative products for similar applications, one or more of which may be used on a particular project
- Editing notes, explaining options and procedures

10.4 Strategies for Developing and Organizing an Office Master Specification

A master guide specification should be developed by an experienced individual or a team of individuals who are thoroughly familiar with specification principles and formats and electronic software capability. The office master specification may be prepared as a separate effort following an established office procedure and drawing from all available resources. The text can be developed in two ways: (1) by compiling and editing sections from previous project specifications and industry association guide specifications or (2) by utilizing commercially available master guide specifications, edited to suit office practice. In planning and preparing an office master specification, carefully analyze the various electronic technologies and methods of editing, storing, and retrieving text for the required sections and for reproducing the final project manual.

10.4.1 Master List of Section Numbers and Titles

When starting the development of an office master specification, first prepare a master list of section numbers and titles needed, along with a scope statement for each section. During development, a comprehensive list facilitates coordination, avoids duplication, and helps prevent omission of required sections. The master list can also be used as a checklist for project specifications. Section numbers and titles should be in accordance with those listed in *MasterFormat*®. The office master list should designate:

- The number and title permanently assigned to each section, with gaps in the numbering of Level 3 sections to allow the addition of other sections needed for a specific project

- A brief description of the work included in each section and related work specified in other sections
- Current status of the development of each section
- Date of the initial preparation and the date of the latest revision of each section
- Name of the individual responsible for preparing and updating each section
- Order of priority for the completion of incomplete or unprepared sections

The more commonly used sections should be prepared first, with other sections postponed until needed for a particular project or available labor permits development. If a commercial master guide specification is being used as the initial basis of the office master specification, the checklist should clearly identify which sections have been edited to suit the practice and which have not, to avoid the assumption by novices that uncustomized sections are acceptable for use.

When setting up the master list of section titles from scratch, as opposed to using the list provided by the commercial master guide specification being used, determine the required scope of each section, on the basis of the work to be covered and the ultimate size (number of pages) of a typical project section. A master guide specification may contain any combination of sections at *MasterFormat®* Level 2 or Level 3. Each firm must make its own judgment on the basis of its practice. The following are some suggestions for scope of master sections:

- A single section may be written covering the entire scope of a division. For example, a single section for a division such as concrete can be useful for small projects.
- Separate sections for Level 2 titles listed in *MasterFormat®* can divide the extent into more manageable pieces.
- If a Level 2 section is too broad for general use because it contains too many types of products or would be unwieldy in size, several Level 3 sections may be written in lieu of the single Level 2 section.
- In some instances, it may be useful to prepare both a Level 2 section and corresponding Level 3 sections for the same scope. The Level 2 section can be used for less complex projects, and the Level 3 sections can be used for more detailed requirements.
- Level 3 sections permit a detailed subdivision of the specifications, making a considerably more modular system. Sections for which decisions have already been made can be started very early in the project without requiring delay until other related product selections have been made.

10.4.2 Standard Formats and Language

Standard section and page formats should be used in the development of a master guide specification, to promote consistency from section to section. The easiest way to do this is to create a master section pro forma or template from which new master guide specification sections or custom project specifications are developed. The pro forma should follow the CSI *SectionFormat™* and *PageFormat™* and should contain the standard articles, paragraphs, and statements used most often, to make it easy to use consistent language throughout the master guide specification system. Refer to Figure 10.1 for an example of a partial section pro forma. Master guide specifications and the master section pro forma should follow the principles of specification language and the use of streamlining for brevity, as discussed in this practice guide.

SECTION [NUMBER]

[SECTION TITLE]

****

SECTION INCLUDES [______________________________].

EDIT ENTIRE MASTER TO SUIT PROJECT REQUIREMENTS: MODIFY AS NECESSARY. DELETE ITEMS WHICH ARE NOT APPLICABLE.

♣ SYMBOL IN LEFT MARGIN INDICATES ACTION IS REQUIRED: EDIT/SELECT/ADD/DELETE TO SUIT PROJECT REQUIREMENTS. INFORMATION BETWEEN BRACKETS [ ] WILL BE DELETED AUTOMATICALLY UNLESS SPECIFIC INDICATION IS MADE TO RETAIN INFORMATION.

****

PART 1—GENERAL

1.01 SUMMARY

♣ A. Section Includes: [______________________________]

****

EDIT PARAGRAPH BELOW TO SUIT PROJECT REQUIREMENTS: ADD SECTIONS AS APPLICABLE.

****

B. Related requirements:

♣ 1. Section [__________] - [___________________________]

♣ 2. Section [__________] - [___________________________]

****

RETAIN ARTICLE BELOW IF APPLICABLE.

****

♣ [1.02 PRICE AND PAYMENT PROCEDURES

1. Allowances: refer to Section 01 21 00 for allowances involving Work of this Section.]

****

ITEMS LISTED UNDER FOLLOWING ARTICLE HAVE BEEN USED WITHIN OTHER PORTIONS OF THIS SECTION. IF PARAGRAPHS CONTAINING THESE REFERENCES ARE DELETED IN PROCESS OF EDITING, DELETE ITEMS HERE ACCORDINGLY.

****

1.03 REFERENCES

A. American Society for Testing and Materials International (ASTM)

♣ 1. [______-______] - [___________________________________]

****

DEPENDING ON SCOPE OF WORK, A PREINSTALLATION CONFERENCE MAY NOT BE NECESSARY. RETAIN ARTICLE BELOW AND EDIT ACCORDINGLY.

****

1.04 ADMINISTRATIVE REQUIREMENTS

A. Pre-installation Meetings

1. Conduct pre-installation meeting in accordance with Division 01 Section 01 31 19.

♣ 2. Convene pre-installation meeting [one week] [__________] prior to commencing work of this Section [prior to purchasing materials for field sample [mock-up]].

****

INCLUDE SUBMITTAL REQUIREMENTS BELOW WHICH ARE CONSISTENT WITH SCOPE OF PROJECT AND EXTENT OF WORK OF THIS SECTION. ONLY REQUEST SUBMITTALS WHICH ARE ABSOLUTELY NECESSARY.

****

1.05 SUBMITTALS

A. General: Submit in accordance with Division 01, Section 01 33 00

B. Product Data:

[Project Name/Number/Date] [Page Number] [Section Title]

Figure 10.1
Example of Partial Section Pro Forma

1. Submit manufacturer's descriptive literature and product specifications for each product.

♣ 2. Include data to indicate [__________________].

C. Shop Drawing:

♣ 1. Indicate typical layout including dimensions and [________________________________].

♣ 2. Submit detail drawings of [__________________].

♣ [3. Submit drawings showing field measured dimensions.]

[4. Submit detail drawings of special accessory components not included in manufacturers product data.]

****

INCLUDE QUALITY ASSURANCE REQUIREMENTS CONSISTENT WITH SIZE AND SCOPE OF PROJECT AND EXTENT OF WORK OF THIS SECTION. EDIT ARTICLE BELOW ACCORDINGLY.

****

1.06 QUALITY ASSURANCE

♣ A. Manufacturer Qualifications: Company specializing in manufacturing products specified in this Section with minimum 5 [_______] years [documented] experience.

♣ B. Fabricator Qualifications: Company specializing in fabricating products specified in this Section with minimum 5 [_________] years [documented] experience.

♣ C. [Installer] [Applicator][Erector] Qualifications: Acceptable to manufacturer with [documented] experience on at least 5 [_____] projects of similar scope in past 5 [____] years.

****

INCLUDE DELIVERY, STORAGE, AND HANDLING REQUIREMENTS CONSISTENT WITH SIZE AND SCOPE OF PROJECT AND EXTENT OF WORK OF THIS SECTION. EDIT ARTICLE BELOW ACCORDINGLY.

****

1.07 DELIVERY, STORAGE, AND HANDLING

A. Comply with requirements of Section 01 65 00 and Section 01 66 00.

****

COORDINATE LENGTH OF WARRANTY PERIOD WITH ADJACENT CONSTRUCTION AND OWNER REQUIREMENTS TO OBTAIN UNIFIED SYSTEM WARRANTY. VERIFY AVAILABLE WARRANTY TIME PERIODS FROM SPECIFIED MANUFACTURERS. EDIT ARTICLE BELOW TO COMPLY WITH PROJECT CONDITIONS AND/OR OWNER'S INSTRUCTIONS. IF STATUTORY ONE-YEAR WARRANTY PERIOD IS SUFFICIENT, DELETE ARTICLE BELOW.

****

1.08 WARRANTY

A. Comply with provisions of Section 01 78 00.

♣ B. Warrant installed units [____________] to be free from defects in material and workmanship for [______] years.

♣ C. Include coverage for [against][_________________].

PART 2—PRODUCTS

****

EDIT LIST OF MANUFACTURERS BELOW TO SUIT PROJECT REQUIREMENTS.

****

2.01 MANUFACTURERS

♣ A. Manufacturers' Names: [________________]

*****OR*****

[Section Title] [Page Number] [Project Name/Number/Date]

(Continued)

2.02 MANUFACTURERS AND PRODUCTS

♣ A. Products and Manufactures: [______________________]

****

RETAIN PARAGRAPH BELOW IF SUBSTITUTIONS WILL BE ALLOWED.

****

♣ [B. Subsitutions: Under provisions of Section 01 25 00.]

****

EDIT DESCRIPTIVE SPECIFICATIONS IN ARTICLE BELOW TO IDENTIFY PROJECT REQUIREMENTS AND TO ELIMINATE ANY CONFLICT WITH SPECIFIED MANUFACTURER'S PRODUCTS.

****

♣ 2.03 MATERIALS [MANUFACTURED UNITS][EQUIPMENT][COMPONENTS]

****

SELECT AND SPECIFY/EDIT COLORS AND FINISHES IN ARTICLE BELOW TO SUIT PROJECT REQUIREMENTS. DELETE INAPPLICABLE ITEMS.

****

♣ 2.04 FINISH [FINISHES][SHOP [FACTORY] FINISHING]

A. [__________] Finish: Color and finish as selected by Architect from sample submitted [____________________].

2.05 ACCESSORIES

PART 3—EXECUTION

3.01 EXAMINATION

A. Examine conditions and proceed with work in accordance with Division 01, Section 01 71 00.

3.02 PREPARATION

♣ 3.03 ERECTION [INSTALLATION][APPLICATION]

♣ A. Install [Erect][Apply][Place] in accordance with manufacturers printed instructions, [State[Municipality] of [______________] standards].

♣ B. Install unit [________] plumb, level, square and free from warp or twist while maintaining dimensional tolerances and alignment with surrounding construction [adjacent surfaces].

3.04 FIELD QUALITY CONTROL

A. General: Comply with requirements of Division 01, Section 01 45 13.

♣ B. Tests: [______________].

♣ C. Inspections: [_____________].

3.05 ADJUSTING

A. Adjust parts for smooth, uniform operation.

3.06 CLEANING

♣ A. Clean as recommended by manufacturer. Do not use materials or methods which may damage finish [surface] or surrounding construction.

END OF SECTION

Figure 10.1
Example of Partial Section Pro Forma (*Continued*)

10.5 Preparing Master Guide Specification Sections

A master guide specification section should cover the items, products, materials, methods, and alternatives that an A/E will encounter most frequently. In addition, each section should conform to standard terminology, designation of options, and appropriate workmanship. The procedures followed for developing a master guide specification section are similar to those discussed earlier for writing specification sections.

The basic steps in preparing a master guide specification section are as follows:

1. Assemble and review resource material from available sources.
2. Determine the specification information necessary for the section. This determination will depend on the type and size of projects anticipated. Judgment and experience must be exercised to prepare master guide specifications adaptable to the various projects and types of construction contracts that will be involved.
3. Confirm that the information designated for the individual section is appropriate and manageable. If the information options are too complex or the scope too broad, consider subdividing the section into two or more sections of narrower scope. Conversely, very simple and uncomplicated information from several related sections may be combined into a single broader scope section.
4. Place information in the section in accordance with the three-part *CSI/CSC SectionFormat*™. A skillfully arranged master guide specification should be relatively easy to read and use.
5. Each article in the three-part *SectionFormat*™ is restricted to a single topic; strict conformance to this principle will make the section easier to read and use.
6. Choices should be arranged so decisions can be made in a logical sequence with no time wasted in editing. It is important that each option be arranged so that it can be identified, easily referenced, modified, or deleted without disturbing other text. It also makes for easier scanning by readers.
7. Determine the method of handling, and identifying optional requirements for each choice. Where numerous options could exist, blank spaces may be provided for information to be inserted. The two main alternatives for multiple-choice options are using brackets or using alternative paragraphs. The decision for handling options usually hinges on how easy it is to read and understand the options, plus the processing technology being used. For instance, mutually exclusive options in individual paragraphs can be more quickly edited by deletion than can options embedded with brackets. However, a computer macro may make the in-line bracketed options just as easy and quick to edit. The A/E should carefully study the ways in which choices are presented in the master guide specification. The information storage and retrieval system used will affect the way in which choices are made available for evaluation and selection or rejection for inclusion in a project specification. Some choices, such as those required by reference standards or brand-name products listed, may be adopted as permanent decisions. Each such decision should be listed separately, identified as such, and reviewed periodically to be sure that the decision is still appropriate.
8. Physically arrange the text according to the *CSI/CSC PageFormat*™.
9. Include notes and instructions to assist the persons editing the section. This information should be clearly identified so it will not be confused with actual specification text. This type of information may be labeled as "EDITING NOTES" or set apart from specification text by some notation, such as indentation, separation, a text box or a line of asterisks above and below the note, a font change, hidden text, or a combination of several schemes. Specification notes affecting an entire article are normally located immediately after the article title and before the affected text. Notes affecting individual paragraphs should normally appear just before the affected paragraph. Notes are used to:

- Provide a brief overview of the content.
- Bring options to the attention of the A/E.
- Identify needed selections and options.
- Provide supplementary information to aid in decision making. This could include commentary on references standards and options within those standards.
- Point out coordination and cross-reference requirements.

Some options needed in project specifications will never be included in the master guide specifications because either there are too many options to consider, such as for color, pattern, and texture, or they are project-specific, such as engineering design data. In addition, other information essential to proper specifying will never appear in the project specifications and therefore is never included in the master text. This information notably includes detailed cost information and actual text of trade association and other referenced standards. The master guide specification system must allow for efficient retrieval of all these types of external information for specific projects. Some of these types of information may be in electronically retrievable form accessible by hyperlinks embedded in the master guide specification, either in the text or in the notes.

10.6 Commercially Available Master Guide Specifications

As of publication date, there are several producers of full-length commercially available master guide specifications based on U.S. products and standards, including American Institute of Architects (AIA)/ARCOM (Architectural Computer Services, Inc.) MASTERSPEC, Construction Sciences Research Foundation (CSRF) SPECTEXT, and Building Systems Design (BSD) SpecLink. Most commercial master specifications are based on *MasterFormat*® section numbers and titles. These products are referred to as master specifications systems. These products are available on a subscription basis and vary significantly from one another in organization, extent, flexibility, and language, style, and automation features. These master specification systems generally operate under commercial and proprietary application software for personal computers. The successful use of a commercial master guide specification by an A/E depends largely on how well it matches the needs and preferences of the firm.

A commercial master guide specification is designed to accommodate a variety of construction types, geographic locations, code jurisdictions, and client types. It is important that a commercial master guide specification be tailored to the particular needs of a firm. Some provisions may be too restrictive or costly to be used for particular applications and should be deleted; inclusion of inapplicable provisions often results in increased cost simply because of the confusion engendered. Failure to delete or edit such provisions could result in unnecessary increases in construction time and cost. Failure to delete inapplicable provisions may become confusing to contractors and others using the documents. If entire specification sections or significant portions of a section are not expected to be used, then they should be deleted when the master is adapted for the firm. In addition, sections not pertaining to the normal work of the firm can be eliminated. When portions are removed, the deleted portions should be clearly identified and remain available for future use if appropriate.

10.7 Updating Master Guide Specification Sections

Updating is an important aspect affecting the effectiveness of a master guide specification. Out-of-date information can cause construction and liability problems and cost time and money. Master guide specifications should be updated on a regular schedule in accordance with an established policy. The commercially available master guide specifications previously listed are updated on differing bases, but typically on a predefined schedule and as needed when problems are identified, or when a significant change has occurred concerning the information contained in a given section. Subscribers to commercial specifications should update their own master guide specification sections as the revised material is received.

The process of updating master guide specifications should include:

- Revising text that has caused problems on a previous project
- Changing text and designations to reflect current referenced standards
- Reviewing choices that have been adopted as permanent
- Reviewing the continuing appropriateness of proprietary, descriptive, and performance-specifying methods and changing to other methods if necessary
- Adding new sections
- Inserting additional options and choices in specification clauses
- Eliminating typographical and other errors
- Updating terminology to be consistent with the CSI formats and office practice

10.8 Additional Considerations

10.8.1 Electronic Technology

Current technology can provide further enhancements for master guide specifications. Use of various types of software and features can provide improved efficiencies and productivity. Such software may involve word processing, spreadsheets, database, computer-aided design (CAD), and building information modeling (BIM). Features can include:

- Notes and instructions that do not require deletion and are maintained as hidden text or tags that do not print (except when desired)
- Features that provide the ability to track or record changes, such as insertions and deletions
- Incorporation of web-based hyperlinks to additional data or resources, such as manufacturer's or reference standard web sites
- Inclusion of embedded objects such as charts, photos, or graphic images
- Shared linkages to other software for CAD, BIM, cost estimating, and scheduling
- Dynamic changes between files, such as needed data, cross referencing, and compilation of listings

- Publishing to a neutral format, such as portable document format (PDF) or internet based posting, for accessibility and usability
- Security features to prevent modification of contract documents by other than those specifically designated

10.8.2 Additional Information

A master guide specification, in itself, is not a complete system. The question of how much additional information is needed depends, in part, on how comprehensive the master guide specification is intended to be. The master guide specification user will still need to refer to product data, project-specific design information, and referenced standards.

10.8.3 Sections on Demand

Usually, each section of a master guide specification to be used on a project is duplicated so that it can be edited. Current files should be readily available in a central location; for example, originals can be kept in a series of notebooks or files stored electronically on a computer network. Stockpiling of paper copies for later use is inefficient because updates can make them obsolete before they can be used. Computer printers and copy machines provide an efficient method of obtaining current printed copies, if required for manual editing.

10.8.4 Review Prior to Distribution

Final project specifications culminate in distribution for pricing. A review prior to distribution can be enhanced by comparing project text to master text or by tracking changes made to the master specification. Copies of project specification files should also be published in a secure medium and shared among team members, by distributing within a computer network, transmitting via electronic mail, or posting to a project web site. National and international companies will have specific decisions and directives relative to the regional and local office practice.

Chapter 11
Shortform Specifications

11.1 Introduction to Shortform Specifications

Some projects of limited scope and extent may need less detailed descriptions than those provided by a typical full length specification. Similarly, for projects with a negotiated contract or design-build delivery method, a specification with reduced detail may be appropriate. In these cases, an architect/engineer (A/E) may choose, or an owner may request, to use a *shortform* specification. Although the name might imply that documents are less than complete, the specification must provide sufficient detail to describe the requirements of the project. It must provide, at minimum, a code-compliant level of information. Unlike the preliminary project descriptions or outline specifications that are part of the early design phases, a shortform specification is a legally enforceable contract document. Shortform specifications should be viewed as the process of logically and consistently editing full-length specifications into a compact form.

11.2 Concept of Shortform Specifications

The purpose of shortform specifications is to provide a concise set of construction specifications commensurate with the size and extent of a project and to reduce the level of detail where it is appropriate. Project specifications can be as brief as a collection of detailed notes on the drawings (as shown in Figure 11.1) or as lengthy as a traditional full-length specification. Depending on the level of detail used, a shortform specification will fall at some point along this continuum of project descriptions. A full-length master guide specification can be condensed into a shortform specification by combining closely related sections and by the judicious removal of requirements unnecessary for projects of limited scope. Correctly developed, the resulting shortform specification will be concise without the omission of critical information. As with full-length specifications, the imprudent removal of language from a specification in order to shorten its length can reduce detail to the point of sacrificing quality control. Thus, the preparation of a shortform specification requires the competence, skill, and technical knowledge of an A/E who is experienced as a specifier and who has the knowledge of construction standards.

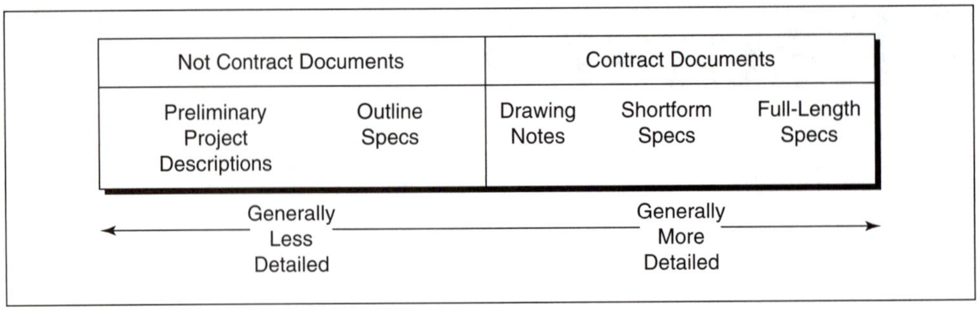

Figure 11.1
Level of Detail

11.3 Characteristics of Shortform Specifications

Shortform specifications are a part of the written portion of the contract documents and share the same characteristics as full-length specifications. These shortform specifications should:

- Express the design intent and a consistent level of project detail despite their short length.
- Include information necessary to achieve the design intent and the required level of detail. Not all requirements, however, can be described briefly, and text should not be condensed solely for the sake of brevity. The level of project requirements expressed by the shortform specifications must be consistent with the level of detail necessary to meet the owner's project requirements.
- Be capable of being priced. Like typical specifications, they should be clear, concise, correct, and complete.
- Be made specific to a project. They should be coordinated with local codes and construction practices.
- Be fully enforceable.
- Parallel the level of drawing detail.

Shortform specifications are made cost-effective by communicating the project requirements in the shortest possible form. They are compact and easy to read, and the language is concise and direct. Because there is less written material to review, coordination and document review are simplified. Less time and cost are usually required to assemble and edit shortform master guide sections.

Shortform specifications might impose limitations. The level of construction detail may be more difficult to maintain because there is less descriptive content. The A/E should carefully determine the amount of detail necessary for each individual specification section. The A/E must be skilled, knowledgeable, and experienced as a specifier to ensure that important items are not omitted and unnecessary detail is not retained. The risk of error might increase as the detail decreases, and the A/E, not the owner, assumes this risk. The A/E and the owner should agree on an acceptable level of requirements for the project and determine the degree of specification detail required accordingly.

11.4 When to Use Shortform Specifications

Shortform specifications might be used for any project, though some types of projects lend themselves more readily than others:

Small Projects Projects of limited extent and cost. The use of a full-length specification may not be consistent with the scope of the project drawings. In such cases, the use of a shortform specification would be justified as long as the desired level of detail is maintained throughout. Similarly, shortform specifications can be used effectively on projects that use standard materials and traditional construction details, as they generally require less descriptive detail in the specifications.

Design-Negotiate-Build Project Delivery Where a contractor has had previous experience with the owner or A/E on a similar project type, there is usually better communication, so the chance of misunderstanding is lessened. The shortform specification can be effective because the contractor is familiar with the level of detail required by the owner, and less descriptive detail may be acceptable in the specifications.

Construction Management Project Delivery Multiple-prime contracts may present coordination complications, and unless the shortform specification is written to cover such situations, full-length specifications are usually more suitable.

Design-Build Project Delivery The level of detail provided by the shortform specifications will depend in large part on the completeness of the owner's project requirements. Shortform specifications can be used effectively in design-build projects because the single contract and single focus of accountability provide many of the same features of the negotiated contract. Because the A/E and the contractor are associated, coordination and communication are improved, which reduces the chance of misunderstanding, and less administrative detail is usually required. The level of detail is purposely reduced to decrease cost and to increase flexibility in product selection, subcontract negotiation, and installation. Items such as shop drawings, samples, testing, inspection, and extended warranties may not be included in the specifications unless required by the owner or by code. The owner will have limited influence, and much of the decision making will be done by the design-build entity.

Owner-Build Project Delivery Because the administrative and installation details are reduced to the minimum, the A/E will have less influence, and more of the decision making will be done by the owner.

Interior Design and Tenant Improvement Projects These projects may involve the installation of standard finishes, furnishings, and equipment. Unless the materials or systems used are unusual, there is usually no need for detailed product descriptions or execution requirements. An acceptable level of detail may be specified by use of reference standards.

Single-Prime Contract Simple projects can work better with shortform specifications because there are generally fewer complex administrative issues.

Projects Where a Shortform Specification Is Specifically Requested The owner should be informed of the reduced detail of the specifications and its limitations.

11.5 Shortform Master Guide Specifications

Shortform master guide specifications are commercially available or may be created by the A/E from a traditional full-length master guide specification. A separate shortform master guide specification may be prepared for each project type commonly designed, such as residential, light commercial, or tenant improvement. The resulting shortform master guide specification can then be efficiently edited to create the project specification. Although a shortform specification may be created by other methods, such as expanding a project description, editing an existing specification, developing a section from industry reference standards, or writing a section from scratch, it is recommended that shortform specifications be developed from a master guide specification. It should be emphasized that editing for brevity and simplification may present a greater chance for the occurrence of errors and omissions. As seen in Figure 11.2, creating project specifications from a

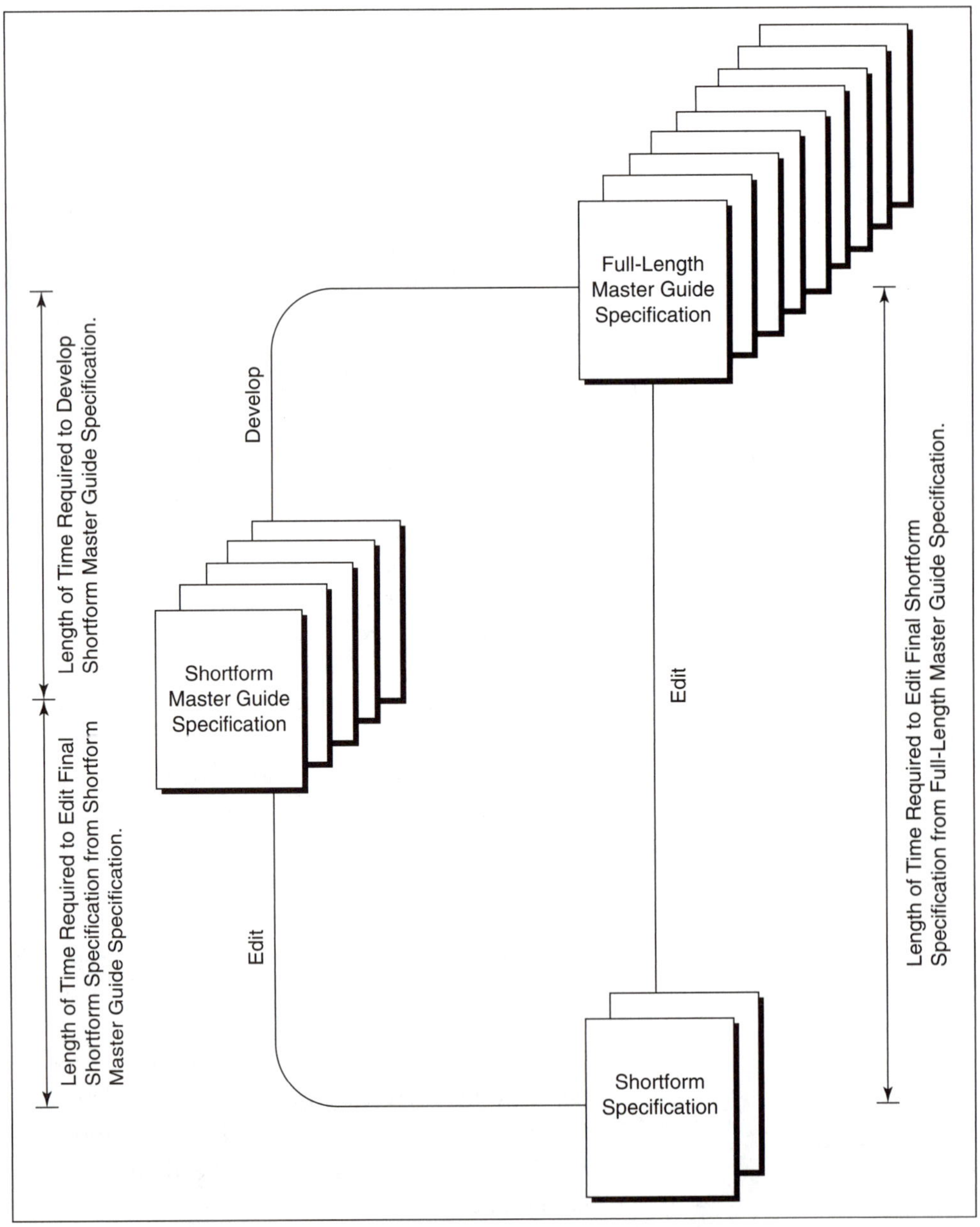

Figure 11.2
Comparison of Time Requirements

shortform master guide specification will usually take less time than editing a traditional full-length specification each time because of the time saved in having a previously created abbreviated master document. The shortform master guide specification will already have been edited and prepared in a condensed format. Although a project may have a unique situation that requires that items be added to the specification, the editing of the shortform master will decrease the likelihood that items will be omitted and the level of project detail will be of the desired level throughout the specification. The following steps should be taken to create a shortform master guide specification for a given project type.

11.5.1 Establish a Specification Type

The A/E should determine the extent of the project types and establish the level of detail the various specification sections must contain. This will help determine which method of specifying is best suited to achieving the required detail. Where the level of extent of a project requires a greater degree of detail, full-length specification sections should be used. The shortform master guide specification for each project type can have combinations of Level 2 and Level 3 *MasterFormat®* section titles. Combinations of master sections may be used within a project manual:

- Full-length sections and shortform sections may be used together in a project manual. The shortform master section list for each project type should be coordinated with the traditional full-length master section list so that full-length and shortform sections can be combined in the final specification without having redundant numbers and titles.
- A broad Level 2 *MasterFormat®* title for each section can be used when less extent is required for a description of the work (e.g., section for Cast-in-Place Concrete may include concrete forms and accessories, reinforcement, and finishing, or a section may include both gypsum plaster and Portland cement plaster).
- Several more focused Level 3 (*MasterFormat®* levels) sections may be created for those instances when a greater level of extent is required than can be provided by a single Level 2 section (e.g., provide Level 2 Precast Concrete; Level 3 Plant-Precast Architectural Concrete and Site-Precast Architectural Concrete).
- A single section may cover all of the titles within a division, and a single section can become the division itself. This can be done when the project is of limited scope (e.g., Division 03—Concrete, which includes all aspects of concrete work, including forms and accessories, reinforcement, and finishing, may be titled "Concrete").

11.5.2 Establish a Master Section List for the Project Type

Determine which sections are applicable to the project type. *MasterFormat®* should be used as the standard organizational system to establish numbers and titles for the sections. The master list should show preferred numbers and titles that have been permanently assigned to each section, with gaps in the numbering to allow for the addition of other project-specific sections. For each section included:

- Provide a description of each section.
- Identify the work included in related sections.

- Provide the current status of each section, including the date of initial preparation and the latest revision.
- Identify the party responsible for preparing and updating the section.
- Determine the order of priority for completion of each section.

11.5.3 Develop a Standard Shortform *SectionFormat™* and *PageFormat™*

A standard section and page format should be developed for a shortform master guide specification. A standard shortform format will present text clearly and at a density best suited for the brief descriptions to be read easily. As indicated by Figure 11.3, the three-part *SectionFormat™* should be used with no more than two levels of paragraphs under each article. A third level may be used occasionally for listing items, but their use should be avoided wherever possible. The use of the three PART headings is recommended because they act as organizational markers to help locate and coordinate information quickly. They serve as a reminder to maintain consistency throughout the project manual. Figure 11.3 illustrates article headings for shortform specifications. The choice of articles and format would be decided by the A/E based on the level of extent required for the particular project. Figure 11.4 illustrates a shortform specification using *SectionFormat™* and *PageFormat™*.

PART 1—GENERAL
1.01 Section includes
1.02 Price and Payment Procedures
1.03 Administrative Requirements
1.04 Action Submittals
    A. Shop Drawings
    B. Product Data
    C. Samples
1.05 Closeout Submittals
1.06 Quality Assurance
1.07 Field [or] Site Conditions
1.08 Warranty

PART 2—PRODUCTS
2.01 Manufacturer
2.02 Description
2.03 Material
2.04 Fabrication
2.05 Finishes
2.06 Accessories

PART 3—EXECUTION
3.01 Examination
3.01 Preparation
3.01 [Erection][Installation][Application]
3.04 Site Quality Control
3.05 Attachments

End of Section
Schedules

Figure 11.3
Example of Shortform Section Article Headings

SECTION 03 30 00

CAST-IN-PLACE CONCRETE

PART 1—GENERAL

1.01 SUMMARY

A. Section includes cast-in-place concrete for interior slabs on grade.

1.02 SUBMITTALS

A. Concrete mix design

B. Concrete delivery slips

PART 2—PRODUCTS

2.01 CONCRETE MATERIALS

A. Portland Cement: ASTM C 150, Type I

B. Normal-Weight Aggregate: ASTM C 33, uniformly graded

C. Water: Potable

D. Reinforcing Bars: ASTM A 615, Grade 60, deformed

E. Welded Wire Fabric: ASTM A 185, welded steel wire fabric, flat sheets

F. Supports for Reinforcement: Comply CRSI specifications

2.02 CONCRETE MIX

A. Comply with ACI 301, ASTM C 94, and ASTM C 116 for normal-weight ready mixed concrete as follows:

1. Compressive Strength (28 Days): 4000 psi (27.6 MPa)
2. Slump: 4 inches (100 mm)

2.03 CONCRETE ACCESSORIES

A. Moisture Retaining Cover: 10 oz. burlap laminated to 4 mil polyethylene sheet.

B. Vapor Retarder: 15-mil high density polyethylene, perm rating not to exceed 0.006 gr/ft^2/hr

PART 3 — EXECUTION

3.01 CONCRETE PLACEMENT AND FINISHING

A. Comply with recommendations in ACI 304R for measuring, mixing, transporting, and placing concrete.

B. Control Joints: Saw-cut joints 1/4 of slab depth not to exceed 10-feet in any direction.

3.02 CONCRETE FINISHING

A. Slab Finish: Trowel finish

B. Flatness/Levelness Requirements:

1. Specified overall Value: FF 30/FL 23
2. Minimum Local Value: FF 25/FL 20

C. Curing Method: Moisture-retaining cover

END OF SECTION

Project name/Project no./date (optional information) 03 30 00-1 Cast-In-Place Concrete

Figure 11.4
Example of Shortform Section

11.6 Writing Shortform Master Guide Specifications

The selected sections of the shortform master guide specification should be written by using streamlining techniques.

11.6.1 Establish Division 01 Requirements

Division 01 requirements should be developed using only those articles that apply to the project type and adding others that are needed to cover situations usually encountered for the project type (e.g., cutting and patching for renovation work). Use Division 01 to specify coordination and quality assurance items. When the requirements of the project allow for less complexity, the sections of Division 01 might be combined into one single section, as shown in Figure 11.5.

11.6.2 Update the Shortform Master Guide Specification Regularly

It is particularly important that the shortform master guide specification be updated at the same time that the traditional full-length master guide specification is updated. This will avoid coordination and consistency errors when sections from both full-length and shortform sections are combined in a project specification.

11.7 Shortform Project Specifications

A shortform specification should be edited from a master guide specification. The master guide specification can be either a traditional full-length master guide specification or a shortform master guide specification. Project specifications may combine both shortform and full-length sections. Performance specifications may also be combined with shortform specifications. Determine which specification sections need to be full length and those that can be shortform. Depending on the size and extent of the project, decide the scope of each section. *MasterFormat*® Level 3 subjects for sections should be combined under a Level 2 title when possible. With any combination, the A/E should decide whether it is convenient or efficient to do so. The use of broad Level 2 sections will reduce the length of specifications by combining similar paragraphs in the section. The A/E should maintain the intent and level of project detail consistently throughout the specification, regardless of the resultant length of the individual sections. The length of a shortform project section can vary from a partial page to several pages, depending on the requirements necessary. The A/E should decide which of the articles shown in *SectionFormat*™ can be omitted and which can be logically combined. Division 01 sections should be coordinated with the procurement requirements, contracting forms, conditions of the contract, and the

	Division 01 General Requirements	Section 01 00 00 General Requirements	
		PART 1—GENERAL	
01 10 00	**Summary**	1.01	Summary of Work
		1.02	Work Restrictions
01 20 00	**Price and Payment Procedures**	1.03	Allowances
		1.04	Alternates
		1.05	Contract Modification Procedures
		1.06	Payment Procedures
01 30 00	**Administrative Requirements**	1.07	Project Management and Coordination
		1.08	Construction Progress Documentation
		1.09	Submittal Procedures
		1.10	Sustainable Design Submittals
01 40 00	**Quality Requirements**	1.11	Quality Assurance
		1.12	Quality Control
		1.13	Sustainable Standards Certifications
01 50 00	**Temporary Facilities and Controls**	1.14	Temporary Utilities
		1.15	Construction Facilities
		1.16	Vehicular Access and Parking
		1.17	Temporary Barriers and Enclosures
01 60 00	**Product Requirements**	1.18	Basic Product Requirements
		1.19	Product Substitution Procedures
		1.20	Product Delivery Requirements
01 70 00	**Execution and Closeout**	1.21	Cleaning and Waste Management
		1.22	Starting and Adjusting
		1.23	Protecting Installed Construction
		1.24	Closeout Procedures
		1.25	Closeout Submittals
		PART 2—PRODUCTS Not Used	
		PART 3—EXECUTION Not Used	

Figure 11.5
Example of Division 01—General Requirements within a Single Section

other specification sections. To avoid repeating information from Division 01, PART 1—GENERAL of each section in Division 02 through 49 should be limited to administrative and procedural requirements that apply only to that portion of the work. Include articles such as submittals and quality assurance only when it is necessary to elaborate further on the generalized requirements already outlined in Division 01. The A/E should be careful not to reduce detail in Division 01 below the level of project requirements; however, because administrative requirements often constitute significant cost items, care should be taken to include only those necessary to the project.

11.8 Techniques for Converting Full-Length Master Guide Specifications to Shortform

Traditional full-length master guide specifications may be edited into a shorter format primarily by deleting detail deemed excessive to secure project requirements. The following techniques will help condense the remaining text.

- Eliminate administrative details in PART 1. Eliminate excessive cross-referencing of related sections. Delete titles of reference standards when the standard is included in other text by abbreviation and number. Delete specific submittal details.
- Combine and edit related sections into a single section (e.g., combine Wood Doors, Plastic Doors, and Steel Doors under one section if the subjects will allow this reduction of detail without compromising understanding).
- Combine and edit related articles or paragraphs, and delete excessive details (e.g., for the preceding door example, it may be possible to combine the hardware item descriptions).
- Combine and edit sentences within an article or paragraph, and delete redundant wording (e.g., for the preceding door hardware example, the description of the finishes might be combined into a single sentence without ambiguity).
- Condense the text by using streamlining techniques; be clear and concise with language, using short imperative or declarative sentences; use lists for materials, manufacturers, and references; and use the colon in place of "shall be" when preparing lists.
- Use reference standards to save text space and set a minimum level of project requirements and workmanship.
- Place schedules on the drawings when it is more practical to do so, rather than reproduce them in the specification. However, in some cases it may be more costly to produce and print schedules on the drawings rather than in the specifications. The U.S. National CAD Standard for Architecture, Engineering, and Construction (NCS) recommends inclusion of schedules as part of the drawing set. NCS does, however, state that schedules developed as electronically generated databases or spreadsheets can be incorporated into the drawing set or specifications. Other schedules are part text and part symbol or diagram that might be best placed on the drawings.
- Follow *SectionFormat*™ as closely as possible, using the least number of paragraph levels.
- Follow *PageFormat*™ as closely as possible, using the least number of indentations.

11.9 Commercially Available Shortform Master Guide Specifications

Shortform master guide specifications are available commercially. These should be scrutinized, not so much for what is covered but for what is not covered, before they are used for a project. Determine the level of detail they represent and the type of project for which they would be appropriate. The guide specifications should then be modified as necessary with the intent of tailoring the sections to the firm's practice and maintaining a consistent level of detail.

Chapter 12
Performance Specifying

12.1 Introduction to Performance Specifying

Performance specifying is a method of indicating characteristics and requirements with measurable properties. Performance specifying can be extended from a single attribute to an entire project, with numerous variations in between. By specifying the end result, performance specifying provides flexibility in optional solutions that all meet the same requirements. Various codes and standards are moving toward more performance criteria.

12.2 Construction Systems and Assemblies

Performance specifying can be used to specify complete systems and assemblies as well as components and construction products. Performance specifying might be applied in this context for a variety of reasons. The owner may need to expedite construction through the use of construction systems and assemblies or may wish to seek the widest range of products and construction methods available. Increased competition for products and use of efficient construction technology may also provide cost savings. The primary advantage of performance specifying lies in the applicability of available products and systems to most traditional project requirements. Utilizing existing technology, performance specifying is practical even for smaller projects and can apply to a complete project or portions of a project. Entire systems and assemblies may require specialized engineering and performance-based criteria. This level of performance specifying may be subject to licensure and regulation by authorities having jurisdiction (AHJs).

12.3 Utilizing Nonstandard Technologies

Many products and systems emerge from research and development before they evolve into standard configurations or solutions. Although they may not be entirely new technologies, their development and use may be at a stage where no alternatives are similar enough to specify descriptively without excluding other variations of that system. Some examples include alternative energy systems, environmental control systems, building monitoring systems, and data and communication systems. When the architect/engineer (A/E) is not familiar with specialized products to describe the best solution, performance specifying may be helpful. As sustainable design and Leadership in Energy and Environmental Design (LEED) accreditation become more commonplace, alternative solutions to the traditional construction are required. Green roofs are an example of nontraditional roofing and require significant reconsideration of traditional roofing requirements.

12.4 Developing New Technology

Where existing products, techniques, or processes do not suit the owner's needs, innovation must be encouraged. Shortcomings in available technology may relate to technical capabilities, construction cost, or expediency. Performance specifying in this context gives the owner the advantage of attracting research-and-development expertise and resources from many elements of the construction industry. New technologies and products are being developed to meet the performance requirements of sustainable and green projects. New technologies include materials consisting of recycled postindustrial and postconsumer waste. These new materials can be utilized by providing performance requirements describing the salient characteristics. There are also disadvantages for the owner involving the time and expense of developing requirements, solicitation, research, design and development, and prototype tests. The requirements must also be able to gain economies of scale of the newly developed technology because considerable expense might be incurred in producing a unique product, system, or assembly. The feasibility of using performance specifications to develop new technology will generally be limited to larger, corporate owners or public agencies with a large volume of construction.

12.5 Delegating Responsibilities

In many instances an owner may not wish to assume full involvement in the design and construction of a facility. Many owners with large, ongoing construction programs maintain architectural or engineering departments but do not assume full design or construction responsibilities. Performance specifying can be utilized by an owner to establish an owner's requirements for a project delivery method such as design-build. Using performance criteria, the owner delegates the tasks of definitive design and construction, yet ensures that needs are met through well-developed performance specifications.

12.6 Roles of Participants

Performance-oriented design, construction documentation, procurement, and contracting alter some of the traditional roles and responsibilities normally assumed by participants in the construction project process.

12.6.1 Owner

The owner's project requirements and the owner's needs are the primary interests. Depending on the project type, the owner may begin the project by defining the project requirements via performance criteria. The performance criteria might then be utilized to prepare a request for proposals using a design-build delivery method. Under other delivery methods, performance specifying may be used to provide flexibility and alternative solutions through traditional construction documents. Using performance criteria, the owner will relinquish some control over the solution. Under this process, the owner must be willing to accept the results of performance criteria. Therefore, to ensure satisfaction with results, the owner must be able to communicate the criteria thoroughly. This feature of performance specifying may encourage the owner to perform a more comprehensive analysis of needs than might be done in a traditional relationship.

12.6.2 Architect/Engineer

Depending on the project delivery method, the A/E might be a part of a design-build entity providing design in response to performance-based owner's project requirements. In the traditional design role, the A/E may determine that particular elements will be performance specified. The A/E must know what levels of performance result in a satisfactory solution and must also be knowledgeable of standards, tests, and methods of verification. Performance specifying may also be a form of design transfer. This is frequently the case with engineered elements such as wood or metal trusses, precast concrete, and certain foundations. Some states may have regulations that limit transfer of design from the A/E of record.

12.6.3 Construction Manager as Adviser

When a construction manager as adviser (CMa) is involved in the project delivery, the CMa might assume a role of coordinating the work of multiple-prime contracts. The CMa might deal with both performance items and the interface of descriptive items with those performance elements. Details of some items may not be known prior to releasing the documents for procurement. Thus, the CMa will be involved in administering procurement of performance items, interfacing these with traditionally designed items, and coordinating with the A/E on matters affecting design.

12.6.4 Contractor

Contractors will have different responsibilities when responding to performance criteria. The contractor might be a part of a design-build entity where an A/E is also part of that entity and involved in providing solutions in response to the owner's performance criteria. The contractor or design-builder must be able to generate solutions within the bounds of performance criteria. This will require design and engineering capabilities and may also require closer liaison with suppliers and manufacturers to acquire supplementary engineering support for particular solutions. If a contractor is bidding or proposing only on a specific system or component as one of multiple-prime contracts, there may be unknown interfaces with other systems. The successful contractor will then be responsible for resolving these details after award of the contract and completion of traditional detailing. Generally, performance-based requirements will not significantly affect labor participation. A possible exception would be the extensive use of prefabricated systems built mostly in the factory rather than on-site. The overall amount of work to be performed, however, is essentially unchanged. Many system, assembly, and component manufacturers will retain erection and installation workers whose work must be coordinated with other systems and assemblies.

12.6.5 Manufacturer

Performance specifications affect manufacturers by placing competing products on the common ground of performance. Manufacturers will be responsible, through contractual lines, for submitting proof of the product's compliance with specified performance criteria. A product's use will be based on performance rather than on a convincing advertising campaign. Manufacturers might be placed in a situation in which they provide additional assistance such as design services. Manufacturers of systems or products may, for example, provide engineering to size their products and meet specific criteria or codes. In some cases the manufacturer might also provide specialized installation.

12.6.6 Authorities Having Jurisdiction

In performance-based procurement, AHJs generally will not have complete construction documents for review at the time of permit application or code reviews. Final documentation is often not completed until some time after the contract is awarded and construction has begun. In such cases, AHJs review the available definitive material as well as the performance specifications for code conformance and, if no conflicts are in evidence, may be asked to issue permits, contingent on inspection and approval of final documentation.

12.7 Determining the Extent

Determining the extent of a performance specification involves two interdependent decisions:

- The elements (pieces, parts, components, assemblies, and systems) to be performance specified
- The extent to which those elements will be performance specified

Although entire structures can be performance specified, it is rarely practical to do so. Some projects have specified only selected subsystems in performance terms, whereas the remainder of the project was found to be more practically handled by other methods of specifying. The extent of performance specifying that best suits the owner's needs should be determined early in the project life cycle.

12.8 Performance Specifying of Project Elements

The first determination is which elements of a project, if any, should be specified in performance terms. Which elements provide a range of available options where competition among those options will provide optimum solutions? The question applies to any scale of project element (i.e., part, component, assembly, or system). For example, a performance specification might permit the use of components of any material that will support specific wind loads. Performance specifying might also permit a choice of load-bearing structural systems. Where a range of options is available, the following conditions favor the use of performance specifications:

- No single, distinct solution is recognized as an exclusive choice in terms of material, configuration, or technique.
- Costs of options are reasonably competitive.
- A system or assembly is a substantial portion of the entire project and does not necessitate a configuration or require such specificity as to preclude options.

Performance specifications may also offer advantages under the following conditions:

- A project element embodies a technology where the state of the art has not yet evolved a standard solution for a given situation.
- Development beyond state of the art is required of an existing product or construction item.
- Nothing exists on the market that will satisfy the owner's design or construction needs.

12.9 Levels of Performance Specifying

Figure 12.1 shows a range of specifying modes from full descriptive to full performance specifications. The broadest levels of performance specifying (H and J) are generally used only in design-build contracts.

The middle levels (D through G) are usually the broadest practical levels that are typically used in traditional construction documents. Functional elements of the project are identified as systems, subsystems, assemblies, or components. Statements of functional requirements are attached to these elements and qualified with technical criteria.

The narrowest levels of performance specifying (B and C) are practical for use in most construction but, in some cases, may be too restrictive to allow the manufacturers to be innovative. Specific components or products are identified and statements of technical criteria are tailored to that type of component. Specifications based on *MasterFormat*® and *UniFormat*™ should not be combined into a single project manual. *MasterFormat*® Section 01 80 00 can be utilized to specify overall performance requirements.

12.9.1 Establishing the Level of Performance Specifying

Establishing the level of performance specifying depends on how much latitude in the results is permitted. Figure 12.1 also illustrates the range of control the A/E may exercise for any given construction item. Greater control over the element's configuration is retained by using a narrow degree of performance specifying. In this case, the particular type of subsystem, component, or item, such as structural-steel frame, integrated light/ceiling system, or aluminum curtain wall, is identified. Performance criteria are based on characteristics intrinsic to that type of component. Latitude in devising a solution within

			Performance/Descriptive Mixture	Sample Text	
Typical Specification Using *MasterFormat*®	Conventional Specifying	A	Total project specified in descriptive terms.	10-gage galvanized annealed steel wire hangers spaced 4 ft.o.c. both ways.	FULL DESCRIPTIVE
	Degree to which sample support requirements can be performance specified	B	Descriptive with performance criteria.	Galvanized annealed steel wire hangers of size and attachment sufficient to support 60 lb. each and spaced 4 ft. 0 in.o.c. both ways.	
		C	Descriptive with overall performance criteria.	Galvanized annealed steel wire hangers of size spacing and attachment sufficient to support a uniform ceiling load of 7.5 lbs/s.f.	
		D	A major assembly specified in *MasterFormat*® as a self-contained component.	Integrated ceiling... capable of supporting based load + 6 lbs/s.f. live load.	
		E	Several major assemblies specified in *MasterFormat*® with interface required.[2]	Integrated ceiling... capable of supporting dead load + 6 lbs/s.f. live load...with key slots along building module lines in order to receive top rail of partitions.	
Design-Build statement of system based on *UniFormat*™	Degree to which systems projects can be performance specified using *PerSpective*®	F	Most major components specified as *UniFormat*™ systems with interface required.	Ceiling/illumination/HVAC component... capable of supporting dead load + 6 lbs/s.f. live load... capable of supporting partitions at any building module line capable of...etc.	
		G	Major components specified as *UniFormat*™ systems to be developed solely according to performance requirements.	Ceiling/illumination/HVAC component... capable of supporting dead load + 6 lbs/s.f. live load... capable of supporting partitions at any building module line capable of...etc.	
		H	Total project(s) specified to be developed as systems solely according to general statements of performance.	Ceiling/illumination/HVAC component... capable of supporting all dead and live loads necessary to structural integrity and the safety of the occupants... capable of receiving partitions universally along a building module grid...capable of...etc.	
	Specifying by Determinates	J	Total project(s) specified according to human requirements alone, requiring translation into suitable design, systems, and technical performance requirements.	...Building occupants shall be provided with adequate light for common office tasks, without harshness or deadening uniformity; attractive ceilings that absorb undesirable noise levels without denying sharpness of voice communication; constructed in such a way that partition may be attached and relocation can be accomplished with disturbance limited to the immediate area and of not more than a day's duration.	FULL PERFORMANCE

Technical Input	Source of Components	Testing/Inspection		
No design input required.	Off-the-shelf products.[1]	Check sample. Observe executed work.	**A**	Typical Specification Using *MasterFormat®* (A–E)
Propose size.		Check sample and proposed size. Observe executed work.	**B**	
Propose, size, spacing, attachment.		Check sample and proposed size, spacing. Observe executed work.	**C**	
Propose many particulars of assembly and interface, depending on degree of performance specifying.	Assembled largely from available products.	Check mockup, standard test data. Observe executed work.	**D**	
	Assembled largely from rationalized available products.		**E**	
Contractor must have design capability and ability to modify or produce components.	Assembled from rationalized available components or from project-designed components.	Design needed tests, test mockups, and prototypes for performance.[3] Check executed work for performance.	**F**	Design-Build statement of system based on *UniFormat*™ (F–J)
Contractor must have highly developed design and manufacturing capability.	Assembled largely from project-designed subsystems.[4]		**G**	
None, until architect/engineer has translated general statements down to the degree of Line G.			**H**	
			J	

(1) Off-the-shelf includes products traditionally fabricated for each project using existing technology.

(2) At about this degree consider switch to *UniFormat*™.

(3) Move testing as much as possible to the factory for intermittent production-line sampling or prototype destructive testing, conducted by national certifying organizations to replace on-site testing and review by local authorities and project architect/engineer. On-site testing will still be needed but at reduced levels.

(4) As first-generation systems develop into second-generation systems over a number of years, more and more project-designed products become off-the-shelf items.

Figure 12.1
Levels of Performance Specifying

the constraints of the element's required performance is inherently allowed. This greater latitude in configuration, material, and process is allowed through a broad degree of performance specifying. Describing subsystems or components in broad terms, such as *exterior closure,* and functional criteria in broad terms, such as *thermal transmission* or *weather resistance,* allows the provider to respond with any of a variety of solutions. Glass curtainwall insulated metal panels, composite panels, precast panels, cementitious panels, or brick veneer may all provide satisfactory and acceptable "exterior walls." Figure 12.2 shows a breakdown of project elements based on *UniFormat*™. (Refer to "Use of Preliminary Project Descriptions" and "Use of Outline Specifications" in Project Delivery Practice Guide and *UniFormat*™ in the appendix for a more detailed discussion.) Such a breakdown of project elements can be used in performance specifications to integrate functionally related items on a subsystem level. Project conditions may sometimes necessitate greater

INTRODUCTION
10 PROJECT DESCRIPTION
1010 Project Summary
1020 Project Program
1030 Project Criteria
1040 Existing Conditions
1050 Owner's Work
1090 Funding
20 OWNER DEVELOPMENT
2010 Site Acquisition
2020 Permits
2030 Professional Services
2050 Other Activities
2080 Budget Project Contingencies
2090 Budget Financing
30 PROCUREMENT REQUIREMENTS
3010 Project Delivery
3020 Solicitation
3030 Instructions for Procurement
3040 Available Project Information
3050 Procurement Forms and Supplements
40 CONTRACTING REQUIREMENTS
4010 Contracting Forms and Supplements
4020 Project Forms
4030 Conditions of the Contract
4040 Revisions, Clarifications, and Modifications

A SUBSTRUCTURE
A10 FOUNDATIONS
A1010 Standard Foundations
A1020 Special Foundations
A20 SUBGRADE ENCLOSURES
A2010 Walls for Subgrade Enclosures
A40 SLABS-ON-GRADE
A4010 Standard Slabs-on-Grade
A4020 Structural Slabs-on-Grade
A4030 Slab Trenches
A4040 Pits and Bases
A4090 Slab-On-Grade Supplementary Components
A60 WATER AND GAS MITIGATION
A6010 Building Subdrainage
A6020 Off-Gassing Mitigation
A90 SUBSTRUCTURE RELATED ACTIVITIES
A9010 Substructure Excavation
A9020 Construction Dewatering
A9030 Excavation Support
A9040 Soil Treatment

B SHELL
B10 SUPERSTRUCTURE
B1010 Floor Construction
B1020 Roof Construction
B1080 Stairs
B20 EXTERIOR VERTICAL ENCLOSURES
B2010 Exterior Walls
B2020 Exterior Windows
B2050 Exterior Doors and Grilles
B2070 Exterior Louvers and Vents
B2080 Exterior Wall Appurtenances
B2090 Exterior Wall Specialties
B30 EXTERIOR HORIZONTAL ENCLOSURES
B3010 Roofing
B3020 Roof Appurtenances
B3040 Traffic Bearing Horizontal Enclosures
B3060 Horizontal Openings
B3080 Overhead Exterior Enclosures

C INTERIORS
C10 INTERIOR CONSTRUCTION
C1010 Interior Partitions
C1020 Interior Windows
C1030 Interior Doors
C1040 Interior Grilles and Gates
C1060 Raised Floor Construction
C1070 Suspended Ceiling Construction
C1090 Interior Specialties
C20 INTERIOR FINISHES
C2010 Wall Finishes
C2020 Interior Fabrications
C2030 Flooring
C2040 Stair Finishes
C2050 Ceiling Finishes
C2090 Interior Finish Schedules

D SERVICES
D10 CONVEYING
D1010 Vertical Conveying Systems
D1030 Horizontal Conveying
D1050 Material Handling
D1080 Operable Access Systems
D20 PLUMBING
D2010 Domestic Water Distribution
D2020 Sanitary Drainage
D2030 Building Support Plumbing Systems
D2050 General Service Compressed-Air
D2060 Process Support Plumbing Systems
D30 HEATING, VENTILATION, AND AIR CONDITIONING (HVAC)
D3010 Facility Fuel Systems
D3020 Heating Systems
D3030 Cooling Systems
D3050 Facility HVAC Distribution Systems
D3060 Ventilation
D3070 Special Purpose HVAC Systems
D40 FIRE PROTECTION
D4010 Fire Suppression
D4030 Fire Protection Specialties
D50 ELECTRICAL
D5010 Facility Power Generation
D5020 Electrical Service and Distribution
D5030 General Purpose Electrical Power
D5040 Lighting
D5080 Miscellaneous Electrical Systems
D60 COMMUNICATIONS
D6010 Data Communications
D6020 Voice Communications
D6030 Audio-Video Communication
D6060 Distributed Communications and Monitoring
D6090 Communications Supplementary Components
D70 ELECTRONIC SAFETY AND SECURITY
D7010 Access Control and Intrusion Detection
D7030 Electronic Surveillance
D7050 Detection and Alarm
D7070 Electronic Monitoring and Control
D7090 Electronic Safety and Security Supplementary Components
D80 INTEGRATED AUTOMATION
D8010 Integrated Automation Facility Controls

E EQUIPMENT AND FURNISHINGS
E10 EQUIPMENT
E1010 Vehicle and Pedestrian Equipment
E1030 Commercial Equipment
E1040 Institutional Equipment
E1060 Residential Equipment
E1070 Entertainment and Recreational Equipment
E1090 Other Equipment

E20	**FURNISHINGS**	**G30**	**LIQUID AND GAS SITE UTILITIES**
E2010	Fixed Furnishings	G3010	Water Utilities
E2050	Movable Furnishings	G3020	Sanitary Sewerage Utilities
		G3030	Storm Drainage Utilities
F SPECIAL CONSTRUCTION AND DEMOLITION		G3050	Site Energy Distribution
F10	**SPECIAL CONSTRUCTION**	G3060	Site Fuel Distribution
F1010	Integrated Construction	G3090	Liquid and Gas Site Utilities Supplementary Components
F1020	Special Structures	**G40**	**ELECTRICAL SITE IMPROVEMENTS**
F1030	Special Function Construction	G4010	Site Electric Distribution Systems
F1050	Special Facility Components	G4050	Site Lighting
F1060	Athletic and Recreational Special Construction	**G50**	**SITE COMMUNICATIONS**
F1080	Special Instrumentation	G5010	Site Communications Systems
F20	**FACILITY REMEDIATION**	**G90**	**MISCELLANEOUS SITE** CONSTRUCTION
F2010	Hazardous Materials Remediation	G9010	Tunnels
F30	**DEMOLITION**		
F3010	Structure Demolition	**Z GENERAL**	
F3030	Selective Demolition	**Z10**	**GENERAL REQUIREMENTS**
F3050	Structure Moving	Z1010	Price and Payment Procedures
		Z1020	Administrative Requirements
G SITEWORK		Z1040	Quality Requirements
G10	**SITE PREPARATION**	Z1050	Temporary Facilities and Controls
G1010	Site Clearing	Z1060	Product Requirements
G1020	Site Elements Demolition	Z1070	Execution and Closeout Requirements
G1030	Site Element Relocations	Z1090	Life Cycle Activities
G1050	Site Remediation	**Z70**	**TAXES, PERMITS, INSURANCE AND BONDS**
G1070	Site Earthwork	Z7010	Taxes
G20	**SITE IMPROVEMENTS**	Z7030	License Fees
G2010	Roadways	Z7050	Permit Costs
G2020	Parking Lots	Z7070	Bond Fees
G2030	Pedestrian Plazas and Walkways	**Z90**	**FEES AND CONTINGENCIES**
G2040	Airfields	Z9010	Overhead
G2050	Athletic, Recreational, and Playfield Areas	Z9030	Profit
G2060	Site Development	Z9050	Construction Contingencies
G2080	Landscaping	Z9090	Financing Costs

Figure 12.2
Levels 1–3 of *UniFormat*™

control over selected elements. For example, under "Exterior Closure," "Curtain Walls" may be further defined as "Metal Curtain Walls" or even "Aluminum Curtain Walls," if necessary. In such cases, the A/E may wish to increase the degree of specificity in the performance criteria to be more consistent with the described component. For example, maximum wind deflection for an "Exterior Closure" might be performance specified as follows:

For exterior walls, at load levels of

1. 0.9 dead + 1 wind (0.9D + 1W),

and

2. 1 dead + live + 0.8 wind (1D + 1L + 0.8W), the deflection due to superimposed load of 1W for (b)1. and 1L + 0.8W for (b)2. should not exceed the following: Recommended Limit: dt = "l"/480 Extreme Limit: dt = "l"/240

The same criterion for "Aluminum Curtain Walls" might be performance specified this way:

> Deflection of any metal framing member in direction normal to wall plane, when tested in accordance with ASTM E 330 Standard Test Method for Structural Performance of Exterior Windows, Curtain Walls, and Doors by Uniform Static Air Pressure Difference, not to exceed 1/175 of clear span, or 3/4-inch, whichever is less, except when plastered surface is affected, in which case deflection not to exceed 1/360 of span.

12.9.2 Response to Performance Criteria

The likely response of the local construction community must also be considered when establishing the extent of performance specifying. Response is a critical factor in performance-oriented procurement. A sufficient number of bidders or proposers must be able to respond to performance criteria and be willing to participate in a performance solicitation. The advantages of performance specifying can be negated by insufficient participation. Initially, bidders or proposers must have the capabilities to respond to the performance-oriented items in the procurement document. These items include administering contracts, designing and engineering performance-specified items, acquiring appropriate systems products, and coordinating installation as necessary. Bidders or proposers may provide necessary services in-house or may retain them on a project-by-project basis as needed. In either case, the extent of performance specifying must be established according to capabilities of the construction community.

12.9.3 Means of Evaluation

The extent of performance specifying will also depend on the means of evaluating the submitted proposals, whether objective or subjective, and constraints imposed by the criteria. A subjective evaluation will allow a broader extent of performance specifying. The ability to accept or reject a proposal or to negotiate particular items might provide greater latitude in performance specifying. If absolute objectivity in evaluation is demanded (a "complies/does not comply" basis), a narrower extent of performance may be best. Broad extent of performance specifying does not necessarily mean longer proposal development time or more involved evaluation.

A contractor who uses systems building components on a regular basis may readily respond to a broader extent of performance specification. However, narrowing the extent of performance specifications may serve only to exclude otherwise acceptable options.

12.10 Resources

Once the extent of performance specifications has been established and the attributes applicable to appropriate project elements have been selected, performance criteria must be established for those attributes. Familiarity with related data such as previous performance specifications, performance standards, industry standards, design guides, and federal regulations is essential in developing performance specifications. Trade associations can also provide standards and data oriented to specific products and materials. The referenced standards or tests used must be performance oriented. Many standards and tests were developed for a specific material under specific conditions and are not applicable to other materials or conditions. Other material-oriented tests, however, may actually have valid applications to other materials and conditions, despite their material-specific origins.

Familiarity with particular standards and tests is important to ensure their validity, applicability, and usefulness in evaluating the end result. Specified standards or tests should reflect project conditions. It serves no purpose to overspecify by requiring compliance with standards having no real utility. Such overspecifying also increases costs by placing an unnecessary burden on the bidder or proposer. Avoiding specification conflicts is critical in performance specifying.

Do not specify the same item in both descriptive and performance terms. This creates a potential conflict between specified performance criteria and actual performance of the described component. Specifying a concrete mix by mix proportions as well as by properties of strength might create a conflict. The bidder or proposer might have the option of complying only with the lesser requirement.

12.11 Drawings-Specifications Coordination

Most performance specifications will be accompanied by drawings that supplement written performance criteria. The drawings may include schematic diagrams, layouts, plans, maximum and minimum dimensions, nominal dimensions, critical dimensions, modular increments, and component arrangements. A broad extent of performance specifying will dictate that these drawings be diagrammatic in composition and content. Conversely, a narrower extent of performance specifying requires more definitive drawings.

Any drawings that represent only a suggested definition and do not require strict adherence to a particular configuration should be so noted. Contradictions between written and graphic material must be avoided. What cannot be communicated verbally should be conveyed graphically, with performance criteria presented in a clear and consistent manner. The principles outlined in this Practice Guide in Chapter 8, "Coordinating Drawings and Specifications," should be followed in performance documents as well.

12.11.1 Extent of Drawings

The level of detail on drawings will depend on the extent of performance specifying and the procurement method being used. Basically, bidders or proposers are provided with definitive drawings that contain blanks where performance-specified items occur. The bidder or proposer fills in the blanks with proposed solutions. A drawing, an outline without detail, or any similar indication may represent the occurrence of a performance-specified element (system or component). However, the A/E must definitively indicate the elements with which the system or component must interface. This information is important to the bidder or proposer. The bidder or proposer will then develop detailed drawings showing proposed performance-specified items and their relationship to other work.

When several major systems are performance specified, definitive site drawings may be developed, whereas structural drawings, plans, elevations, sections, and details allow for completion by the successful bidder or proposer. Definitive plans may sometimes be drawn, but with only nominal dimensions indicated where alternative solutions might result in different dimensions. Likewise, foundation drawings and structural layout drawings must accommodate the alternatives allowed by the performance specifications.

Where materials or configurations are optional, they should be drawn only as generic or schematic representations. Elevations and sections may illustrate the arrangement of elements, but only nominal dimensions and abstract material representations should be indicated wherever these vary according to the proposed solutions. Details of definitively described items should be drawn indicating interface with generically illustrated, performance-specified items.

The successful bidder or proposer will complete the drawings, indicating proposed solutions to performance-specified items and their interfaces with the traditionally designed elements drawn by the A/E. A wide range of performance specifying will allow the bidders or proposers to develop a proposal for most, if not all, elements of the project.

The construction drawings will consist of site drawings showing the project outline and design drawings showing schematic plans, elevations, and sections. The successful bidder or proposer will develop definitive drawings for the project.

If the bidder or proposer is to develop the design, the construction drawings might include only a definitive location plan. Because the project's final design is not yet determined, a blank area, or *footprint,* in which the project will be sited, should be indicated on the site plan.

The design criteria and schematic drawings can be included as part of the performance specifications. The successful bidder or proposer will then develop documentation to substantiate the solution. At the broadest extent of performance specifying, such as a turnkey or design-build project, drawings will consist only of existing condition drawings.

Design guidance should be communicated primarily through design criteria included in the performance requirements, supplemented only with schematic drawings. The successful bidder or proposer will develop construction drawings for the project.

If multiple-prime contracts will be awarded for several performance-specified components or systems, each bidder or proposer may offer a solution for only one particular system. The A/E provides definitive construction drawings for all descriptive items, indicating only generically the performance-specified items. The successful bidder or proposer will develop a solution for the performance-based system, indicating its interfaces with the descriptive items. Where multiple performance-based elements interface with each other, those details will have to be coordinated after each interfacing system has been selected. If a prime contract will be awarded for each of several systems assembled prior to delivery, there will be no definitive drawings accompanying the performance specifications. The A/E provides only a schematic project design to each bidder or proposer. These drawings should provide enough information to determine configuration, details, and quantities of the system, thus enabling development of a solution. Bidders or proposers will submit definitive drawings for their respective system.

Not until final selection of each system has been made will the A/E begin to develop definitive construction drawings, integrating performance-specified items with each other and with descriptive items.

12.12 Performance Specification Format

The three-part *SectionFormat™* should be used in performance specifying to maintain consistency and flexibility in the development of a project manual. Figure 12.3 shows *SectionFormat™* as it might be applied to a performance specification. Some article headings have not been listed because they typically would not apply to performance specifications. One additional broad article title has been added as well as several narrow article titles. This is in keeping with the concept that all article titles are optional and should be carefully selected to meet the exact requirements of the specification section. The detailed descriptions of the articles are presented in *SectionFormat™* and are generally applicable to performance specifications as well. The following discussion relates to the special circumstances of performance specifying; Figure 12.4 is a sample performance specification.

12.12.1 PART 1—GENERAL

Most of the administrative and procedural requirements used in PART 1—GENERAL are possible criteria for a performance specification. The article "System Description" should be used to define the interface of the system being specified, with systems or

PART 1—GENERAL

SUMMARY
- Section includes
- Products Supplied but Not Installed under This Section
- Products Installed but Not Supplied under This Section
- Related Requirements

PRICE AND PAYMENT PROCEDURE
- Allowances
- Unit Prices
- Alternates

REFERENCES
- Definitions

ADMINISTRATIVE REQUIREMENTS
- Coordination
- Preinstallation Meetings
- Sequencing
- Scheduling

SUBMITTALS
- Prior to Bid
- With Bid
- Prior to Fabrication
- Prior to Installation
- Prior to Final Acceptance

MAINTENANCE MATERIALS SUBMITTALS
- Extra Materials
- Maintenance Service

QUALITY ASSURANCE
- Qualifications
- Manufacturer Qualifications
- Bidder Qualifications
- Designer Qualifications
- Installer Qualifications
- Testing Agency Qualifications

DELIVERY, STORAGE, AND HANDLING

PROJECT/SITE CONDITIONS
- Environmental Requirements
- Existing Requirements
- Field Measurements

WARRANTY
- Manufacturer's Warranty
- Special Warranty

PART 2—PRODUCTS

PERFORMANCE
- Attributes
- Requirements
- Criteria
- Tests

COMPONENTS
- Attributes
- Requirements
- Criteria
- Tests

FINISHES

ACCESSORIES

SOURCE QUALITY CONTROL

PART 3—EXECUTION

EXAMINATION

PREPARATION
- Protection of Existing Surfaces
- Surface Preparation

ERECTION/INSTALLATION/APPLICATION

FIELD QUALITY CONTROL

ADJUSTING

CLEANING

PROTECTION

END OF SECTION

SCHEDULES

Figure 12.3 SectionFormat as Applied to Performance Specifications

products specified in other sections. Performance requirements in PART 1 should include those applicable to an assembly specified in several sections, such as an assembly with a fire-resistance rating. This article should not be used for stating the desired attributes of the specified products, because this properly belongs in PART 2—PRODUCTS. Submittals are an important part of the performance specification concept. The types of submittals required for performance verification include shop drawings, product data, samples, test reports, certificates, and design data. Some submittals may require review and approval, whereas others are forms of documentary evidence or for information only. These submittals can be separated into quality assurance and quality control submittals, with provisions to ensure correct responsibility for design liability. Submittals

SECTION 06 10 00 ROUGH CARPENTRY

PART 1—GENERAL

1.1 SUMMARY

A. Section includes:

1. Wood grounds, nailers, furring, blocking, and sleepers
2. Plywood backer panels

1.2 REFERENCES

A. Lumber standard: Comply with PS-20 and with applicable rules of the respective grading and inspecting agencies for species and products indicated.

B. Plywood product standards: Comply with PS 1 (ANSI A199.1) or, for products not manufactured under PS 1 provisions, with applicable APA Performance Standard PRP-108 for type of panel indicated.

1.3 DEFINITIONS

A. Rough carpentry includes carpentry work not specified as part of other Sections and generally not exposed, unless otherwise specified.

1.4 SUBMITTALS

A. Product Data:

1. Chemical treatment manufacturer's instructions for handling, storing, installation, and finishing of treated material.
2. Treating plant's certification of compliance stating type of preservative used and method of treatment employed, net amount of preservative retained, and compliance with applicable standards.
3. For waterborne treated products, include statement that moisture content of treated materials was reduced to levels indicated prior to shipment to project site.

B. Certification that chemical treatment complies with specification for each type of treatment.

1.5 QUALITY ASSURANCE

A. Ensure all preservative is adequately fixed in wood. Reject lumber with surface residues of white salts. Provide wood that is kiln-dried after treatment or prefinished with a sealer.

B. Obtain approvals from Building Official for alternative wood preservative treatment.

C. No products used within the interior of the building shall contain urea formaldehyde glue.

1.6 FIELD CONDITIONS

A. Examine substrates and supporting structure and the conditions under which work is to be installed. Do not proceed with the installation until unsatisfactory conditions have been corrected.

1.7 DELIVERY, STORAGE, AND HANDLING

A. Keep materials under cover and dry. Protect against exposure to weather and contact with damp or wet surfaces. Stack lumber as well as plywood and other panels; provide for air circulation within and around stacks and under temporary coverings including polyethylene and similar materials.

1. For pressure-treated lumber and plywood, place spacers between each bundle to provide air circulation.

PART 2—PRODUCTS

2.1 LUMBER, GENERAL

A. Lumber standards: Furnish lumber manufactured to comply with PS 20 "AmericanSoftwood Lumber Standard" and with applicable grading rules of inspection agencies certified by American Lumber Standards Committee's (ALSC) Board of Review.

B. Inspection agencies: SPIB - Southern Pine Inspection Bureau.

C. Grade Stamps: Provide lumber with each piece factory-marked with grade stamp of inspection agency evidencing compliance with grading rule requirements and identifying grading agency, grade, species, moisture content at time of surfacing, and mill.

D. Nominal sizes are indicated, except as shown by detail dimensions. Provide actual sizes as required by PS 20, for moisture content specified for each use.

1. Provide dressed lumber, S4S, unless otherwise indicated.
2. Provide seasoned lumber with 19 percent maximum moisture content at time of dressing and shipment for sizes 2 inches or less in nominal thickness, unless otherwise indicated.

(project number) ROUGH CARPENTRY 06 10 00

3. "Standard" grade.
4. Southern Pine graded under SPIB rules.

2.2 WOOD GROUNDS, NAILERS, BLOCKING, AND SLEEPERS

A. Provide lumber for support or attachment of other construction including rooftop equipment curbs and support bases, cant strips, bucks, nailers, blocking, furring, grounds, stripping, and similar members.
B. Fabricate miscellaneous lumber from dimension lumber of sizes indicated and into shapes shown.
C. Grade: "Standard" grade light-framing-size lumber of any species or board-size lumber as required. No. 2 boards per SPIB rules.
D. Wood grounds, nailers, and sleepers shall be pressure treated as specified herein.

2.3 PLYWOOD PANELS

A. Construction panel standards: Comply with PS 1 "U.S. Product Standard for Construction and Industrial Plywood" for plywood construction panels and, for products not manufactured under PS 1 provisions, with APA PRP-108.
B. Trademark: Furnish construction panels that are each factory-marked with APA trademark evidencing compliance with grade requirements.
C. Electrical or telephone equipment backing panels: DOC PS-1, exposure 1 CD plugged, fire-retardant treated, Thickness: Minimum 15/32 inch. Painting requirements per work described in Division 26.

2.4 FASTENERS

A. Provide fasteners of size and type indicated that comply with requirements specified in this article for material and manufacture.
B. Nails, wire, brads, and staples: FS FF-N-105.
C. Power driven fasteners: National Evaluation Report NER-272.
D. Wood screws: ANSI B18.6.1.
E. Lag bolts: ANSI B18.2.1.
F. Bolts: Steel bolts complying with ASTM A 307, Grade A; with ASTM A 563 hex nuts and, where indicated, flat washers.

2.5 MISCELLANEOUS MATERIALS

A. Water-repellent preservative: NWWDA-tested and -accepted formulation containing 3-iodo-2-propenyl butyl carbonate (IPBC) as its active ingredient.

2.6 PRESSURE TREATMENT OF WOOD

A. Preservative treatment
   1. ACQ—Ammoniacal copper quarternary compound: Pressure-injected.
   2. Use 0.25 lb/cu ft retention.
   3. Kiln dry after treatment to 19 percent maximum moisture content for lumber and 18 percent for plywood.
   4. Optional preservative treatment: CDDC: Copper hydroxide sodium dimethyldithiocarbamate
B. Fire Retardant Treatment (for plywood backing panels only): AWPA C27 Type A.

PART 3—EXECUTION

3.1 INSTALLATION, GENERAL

A. Discard units of material with defects that impair quality of rough carpentry construction and that are too small to use in fabricating rough carpentry with minimum joints or optimum joint arrangement.
B. Set rough carpentry to required levels and lines, with members plumb and true to line and cut and fitted.
C. Fit rough carpentry to other construction; scribe and cope as required for accurate fit. Coordinate location of furring, nailers, blocking, grounds, and similar supports to allow attachment of other construction.
D. Securely attach rough carpentry work to substrate by anchoring and fastening as indicated.
E. Use screws, unless otherwise indicated. Select fasteners of size that will not penetrate members where opposite side will be exposed to view or will receive finish materials. Make tight connections between members. Install fasteners without splitting of wood; predrill as required.
F. Use IPBC-treated products at interior locations and ACO- or CDDC-treated products at exterior locations.

(project number) ROUGH CARPENTRY 06 10 00

(*Continued*)

3.2 INSTALLATION OF WOOD GROUNDS, NAILERS, BLOCKING, AND SLEEPERS

A. Install wood grounds, nailers, blocking, and sleepers where shown and where required for screeding or attachment of other work. Form to shapes as shown and cut as required for true line and level of work to be attached.

B. Attach to substrates as required to support applied loading. Countersink bolts and nuts flush with surfaces, unless otherwise indicated. Build into masonry during installation of masonry work. Where possible, anchor to formwork before concrete placement.

C. Install permanent grounds of dressed, preservative treated, key-beveled lumber not less than 1-1/2 inches wide and of thickness required to bring face of ground to exact thickness of finish material involved. Remove temporary grounds when no longer required.

3.3 INSTALLATION OF CONSTRUCTION PANELS

A. Comply with applicable recommendations contained in Form No. E30, "APA Design/Construction Guide—Residential & Commercial," for types of construction panels and applications indicated.

B. Fastening: Plywood backing panels: Nail, bolt, or screw to supports.

END OF SECTION

(project number) ROUGH CARPENTRY 06 10 00

Figure 12.4
Sample Performance Specification

may be requested at several points in time: prior to bid or proposal, with bid or proposal, prior to fabrication, prior to installation, and prior to final acceptance. Under the broad article "Quality Assurance," the qualification requirements must include credentials for entities involved, such as manufacturer, contractor, designer, installer, and testing agency. These credentials may require engineering expertise and licensure by AHJs.

12.12.2 PART 2—PRODUCTS

The generic term *products* refers to elements including systems, assemblies, and components. As the end results rather than the means of achieving the end are specified, the article titles "Manufacturers," "Materials," "Manufactured Units," "Equipment," "Accessories," "Mixes," and "Fabrication" are seldom used in a performance specification. This type of information might typically be determined by the manufacturer, fabricator, supplier, or contractor. The added article "Performance" is used specifically to describe the design requirements and desired performance. This is accomplished by listing the performance attributes of the system. For each attribute, the A/E must select applicable requirements stating discrete needs or expected results. These requirements must be limited by criteria statements that set limits or give standards. Related to each criterion will be a test by which the performance will be evaluated. Common criteria tests are:

- Review of drawings
- Review of design calculations
- Review of computer simulations
- Independent testing laboratory certification
- Supplier certification
- Tests of mock-ups and prototypes
- Field inspection

For each attribute there may be more than one requirement and, similarly, there may be several criteria for a single requirement. Also, more than one test may be applicable for a single criterion. A useful reference for the specifier is Figure 12.5, which lists various attributes along with sample requirements, criteria, and tests. The attributes are grouped into the following categories:

- Safety and protection
- Functional
- Sensible
- Practical

In more complex systems, it may be necessary to define the attributes of major components, as well as those of the total system. This subdivision of attributes will provide a better-organized document reflecting more detailed information and will make the specification easier to understand and respond to with a solution. "Components" is a generic article title used to list the performance attributes of the components of a system or assembly. The actual name of the component may be used as the article title. Because a system or assembly is composed of more than one component, there can be several component articles. For example, "Suspension Grid," "Acoustical Panels," and "Luminaires" are all component articles that might be performance specified as part of an integrated ceiling system. For each component, appropriate attributes, requirements, criteria, and tests must be specified. When using performance specifications, the A/E must be aware of the state of technology involved in the systems and components being specified.

A requirement to advance the state of the art may produce a limited response unless there is enough volume of work to support the research-and-development costs. If the A/E is looking for the most cost-effective solution using existing technology, care must be exercised not to over specify.

12.12.3 PART 3—EXECUTION

This part of a performance specification section covers execution provisions pertinent to the systems and components specified in PART 2—PRODUCTS. The articles typically found under PART 3—EXECUTION of *SectionFormat™* are generally applicable for performance specifying as well. The article "Erection/Installation/Application" may or may not be used to describe the installation, depending on the extent of the system being specified. The installation or erection may be dependent on the actual product developed by a manufacturer for the project.

However, several products may meet the performance specification, each with its own application, or the installer may be allowed installation options. In some situations, only a statement requiring compliance with manufacturer's installation instructions is necessary. Extra care should be taken to avoid conflicts caused by specifying both the final result desired as a performance requirement in PART 2 and a method of achieving the requirement in PART 3. Division 01—General Requirements includes Section 01 82 00—System Performance Requirements, along with numerous Level 2 and 3 sections for various types of systems. This system performance may establish performance requirements for a result, without reference to specific products. This may be applicable to a design-build project delivery and indicate requirements that may be achieved using alternative products, solutions, or technologies that are not specified in Divisions 02 through 49.

Attribute	*Requirement (R)/Criteria (C)*	*Test*
HEADING I: SAFETY AND PROTECTION		
11 Fire Safety		
(01) Fire Areas	R: Control fire hazard neighboring structures.	
	C: Limit distance between structures: limit area within fire barriers; limit ceiling height.	1
(02) Fire Barriers	R: Control the spread of fire.	
	C: Require fire walls, fire stops, fire-resistance separation between egress openings; require that barrier penetrations maintain rated fire endurance; require fire dampers.	1
(03) Egress Means	R: Provide means for emergency evacuation.	
	C: State minimum number of exits, maximum travel distance to exits and other means of egress; require minimum width for public corridors and public stairways; limit obstruction by door swing or equipment installation; require exit signs.	1
(04) Protective Devices	R: Provide warning devices and automatic fire extinguishing equipment.	
	C: State conditions under which automatic fire detection systems, smoke detection systems, sprinkler systems, extinguishing systems, or other protection devices should be provided.	1.5
(05) Fire Resistance/Combustibility	R: Maintain integrity for sufficient time to permit evacuation or control of fire.	
	C: Require use of noncombustible materials, state minimum hours of fire resistance or classification.	1.3
(06) Fire Load/Fuel Contribution	R: Control fuel contribution of materials.	
	C: State maximum potential heat (contribution to fire load) in BTU per hour or BTU per square foot of material.	3
(07) Surface Spread of Flame	R: Control surface spread of flame.	
	C: State maximum flame spread or flammability of rating.	3
(08) Flame Propagation	R: Control propagation of flame through enclosed spaces.	
	C: State maximum flame propagation index.	1,3
(09) Smoke Generation	R: Control amount and toxic effect of smoke produced.	
	C: State maximum smoke development rating; state maximum optical density and maximum time to reach critical density; limit toxicity of smoke, require that smoke be non-noxious.	3
(10) Smoke Propagation	R: Control propagation of smoke through enclosed spaces.	
	C: Require smoke-tight joints, provide for venting of smoke areas.	1,3
(11) Accidental Ignition	R: Protect against accidental ignition of fire.	
	C: Design to prevent spark formation; limit equipment overheating; require equipment mounting to permit adequate ventilation.	1
12 Life Safety (other than fire)		
(01) Physical Safety	R: Protect against physical hazards.	
	C: Require guardrails, handrails, protective covers on moving parts; slip-resistant surfaces.	1
(02) Electrical Safety	R: Protect against electrical hazards.	
	C: Require protective cover, insulation, and grounding: require safety controls and interlocks.	1
(03) Toxicity	R: Control dangerous materials and substances.	
	C: Limit toxicity of materials, surfaces, and finishes; limit toxic emissions below stated temperatures; limit toxic venting and leakage.	3
(04) Chemical Safety	R: Protect against hazard from chemical substances.	
	C: Identify chemicals and agents, including concentration and anticipated frequency of use, to which the system will be exposed; indicate the level of atmospheric pollution permitted.	1

Test Codes:
1. Design Drawings 2. Design Calculations 3. Laboratory Certification 4. Prototype Testing 5. Inspection

Attribute	*Requirement (R)/Criteria (C)*	*Test*
(05) Azoic Protection	R: Protect against infection from biological sources. C: Identify insects, vermin, fungi, microorganisms, and other biological contaminants likely to be cncountered and state level of protection to be provided.	1
13 Property Protection		
(01) Theft Security	R: Protect equipment and contents against theft. C: Design to control unauthorized entry and access.	1
(02) Security against Vandalism	R: Protect against malicious damage. C: Design to resist malicious damage.	1.5
(03) Resistance to Misuse	R: Protect against accidental or deliberate misuse. C: Design to prevent improper usage. Design for failsafe operation. Perform factory adjustment. Provide instructions.	1,5
14 Accessibility Considerations		
(01) Physical Access	R: Provide for physical access by impaired individuals. C: Design to provide at least one means of ingress and egress for individuals in wheelchairs.	1
(02) Mobility-Impaired Usage	R: Provide for building usage by mobility-impaired individuals, if appropriate. C: Design to permit mobility-impaired individuals access to and use of facilities and equipment, such as restrooms, drinking fountains, vending machines, elevators.	1
(03) Vision-Impaired Usage	R: Provide for building usage by vision-impaird individuals, if appropriate. C: Design to permit vision-impaired individuals access to and use of facilities and equipment, such as restrooms, drinking fountains, vending machines, elevators.	1
(04) Hearing-Impaired Usage	R: Provide for building usage by persons with hearing deficiencies, if appropriate. C: Design to permit hearing-impaired full usage of building services, such as, fire alarm systems, door bells, and audible signals system.	1
HEADING 2: FUNCTIONAL		
21 Strength		
(01) Static Loading	R: Sustain gravity loads and superimposed and specified vertical and lateral loads. C: State dead loads to be supported, including forces transmitted from other systems. Specify how and where loads shall be transmitted from other systems.	2
(02) Live Loading	R: Sustain dynamic loads. C: Describe live loads to be supported, including snow load. Identify concentrated loads and state design floor loads.	2
(03) Horizontal Loading	R: Sustain wind loads and other lateral loads. C: For exterior walls, state design wind speeds and other live loads. State typhoon or hurricane conditions. For partitions, state lateral design load per square foot of partition area.	2
(04) Deflection	R: Limit deflection. C: State maximum acceptable deflections.	2
(05) Thermal Loading	R: Sustain loads due to temperature change. C: State the temperature extremes to be used for design.	2
(06) Structural Serviceability	R: Retain serviceability under load and deflection. C: Require structure to sustain design loads without causing local damage.	2
(07) Seismic Loading	R: Sustain earthquake loads. C: State the seismic zone to be used for design.	2

Test Codes:
1. Design Drawings 2. Design Calculations 3. Laboratory Certification 4. Prototype Testing 5. Inspection

(Continued)

Attribute	*Requirement (R)/Criteria (C)*	*Test*
(08) Impact Loading	R: Sustain impact loads and forces. C: Describe the source and magnitude of any impact loads to be sustained.	2
(09) Penetration Resistance	R: Protect against damage from concentrated loads. C: Describe magnitude and location of concentrated loads.	2
22 Durability		
(01) Impact Resistance	R: Resist surface degradation due to point impact. C: Limit surface indentation due to specified impact load.	3
(02) Moisture Resistance	R: Resist degradation when exposed to water or water vapor. C: Design for use in specified range of humidity. Limit permanent effect to exposure to water, water retention, and absorption.	3
(03) Thermal Resistance	R: Resist degradation when exposed to temperature ranges expected in normal use. C: Limit physical change when exposed to specified temperature range.	3
(04) Corrosion Resistance	R: Resist degradation when exposed to corrosive agents. C: Limit corrosive effect observed after specified exposure to salt spray or fog; require corrosive-resistant surface treatment: design to avoid contact of dissimilar metals.	3
(05) Chemical Resistance	R: Resist degradation when exposed to chemicals. Resist staining or damage from soluble and insoluble salts, alkali attack, and oxidation. C: Limit changes in appearance or other specified property after exposure to specified chemicals.	3
(06) Weather Resistance	R: Resist degradation when exposure to specified period of simulated weathering. C: Limit changes observed after exposure to specified period of simulated weathering.	3
(07) Ultraviolet Resistance	R: Resist degradation due to exposure to ultraviolet light. C: Limit discoloration after ultraviolet exposure.	3
(08) Surface Serviceability	R: Resist cracking, spalling, crazing, blistering, delaminating, chalking, and fading. C: Limit surface changes observed after exposure to simulated conditions of use.	3
(09) Stain Resistance	R: Resist permanent discoloration when exposed to staining agents and chemicals. C: Limit visual evidence of permanent stains due to treatment with identified agents.	3
(10) Absorbency	R: Resist tendency to absorb and retain water. C: Limit quantity of water retained after specified exposure.	3
(11) Cleanability	R: Resist damage from routine maintenance and cleaning; permit removal of identified stains. C: Limit discoloration or surface change after simulated cleaning with specified cleaning agents.	3
(12) Color Resistance	R: Resist fading over time. C: Limit discoloration after stared period.	3
(13) Friability/Frangibility	R: Resist crumbling and brittle fracture. C: Limit damage observed after specified loading.	3
(14) Abrasion Resistance	R: Resist degradation due to rubbing. C: Limit weight loss after specified number of abrasion cycles.	3
(15) Scratch Resistance	R: Resist degradation due to scratching. C: Limit rating on pencil hardness scratch scale.	3
(16) Dimensional Stability	R: Control dimensional changes resulting from changes in environment. C: Limit volume change and movement under specified exposure to moisture and temperature variation.	3
(17) Cohesiveness/Adhesiveness	R: Resist peeling and delamination. C: Limit peeling or delamination failures under specified simulated loading.	3

Test Codes:
1. Design Drawings 2. Design Calculations 3. Laboratory Certification 4. Prototype Testing 5. Inspection

Attribute	Requirement (R)/Criteria (C)	Test
(18) System Life	R: Function properly for identified period. C: Limit failure under accelerated life test. Design life of components consistent with specified life of system.	3,4
23 Transmission Characteristics		
(01) Heat	R: Control heat transmission. C: Design for specified Thermal Transmittance ("U" value).	2
(02) Light	R: Control light transmission. C: Design for specified percentage of light or radiation transmission.	2
(03) Air Infiltration	R: Resist leakage of air. C: Limit infiltration under specified pressure or wind load. Design for specified maximum leakage.	2,3,4
(04) Vapor Penetration	R: Resist vapor penetration. C: Design vapor barrier for minimum vapor permeability.	2,3,4
(05) Water Leakage	R: Resist water leakage. C: Limit infiltration under specified pressure or wind load design for specified maximum leakage.	2,3,4
(06) Condensation	R: Control admission and condensation of moisture. C: Design to provide moisture barriers and thermal breaks.	2,4
24 Waste Products and Discharge		
(01) Solid Waste	R: Control production of solid waste. Provide for elimination or emission and prevent undesired accumulation. C: Design to accommodate waste produced or accumulated. Require identification of wastes produced.	1,2
HEADING 3: SENSIBLE		
(02) Liquid Waste	R: Control production of liquid waste. Provide for elimination or emission and prevent undesired accumulation. C: Design to accommodate waste levels produced, accumulated or omitted. Require identification of waste produced.	2
(03) Gaseous Waste	R: Control production of gases. Provide for elimination and prevent undesired accumulation. C: Design to accommodate levels of gas accumulated or emitted. Require identification of gaseous waste emitted.	1,2
(04) Odor	R: Control formation and persistence of odors. C: Design to prevent odor formation.	1
(05) Particulate Discharge	R: Control production of particulate wastes. Provide for collection of waste and prevent undesired accumulation. C: Design to accommodate amount of particulate waste produced. Limit particulate concentration.	1,2
(06) Thermal Discharge	R: Limit of thermal energy and vibration. Provide for control or reabsorption. C: Design to control thermal discharge produced below specified levels.	2
(07) Radiation	R: Limit emission of radiation. Provide for control or reabsorption. C: Design to control radiation discharge produced below specified levels.	2
25 Operational Characteristics		
(01) Method of Operation	R: Provide operating methods consistent with function. C: List desired operating modes.	1,2,3,4
(02) Results of Operation	R: Provide output consistent with function. C: List desired output quantities and rates.	1,2,5
(03) Cycle Time/Speed of Operation	R: Provide cycle times to accommodate functional requirements. C: List desired repetition rates.	1,2,5

Test Codes:
1. Design Drawings 2. Design Calculations 3. Laboratory Certification 4. Prototype Testing 5. Inspection

(*Continued*)

Attribute	Requirement (R)/Criteria (C)	Test
31 Aesthetic Properties		
(01) Arrangement	R: Provide order, organization, or relationship appealing to visual perception. C: Design for pleasing relationships between elements and components.	1
(02) Composition	R: Provide unified appearance appealing to visual perception. C: Design for pleasing overall appearance.	1
(03) Texture	R: Provide surface finishes appealing to tactile perception. C: Design surface finishes pleasant to touch and feel.	1,4,5
(04) Color/Gloss	R: Provide finishes with pattern or luster appealing to visual perception. C: Design surface finishes for pleasing appearance.	1,4,5
(05) Uniformity/Variety	R: Provide appropriate consistency or variety of visual environment. C: Design to provide pleasing variety of colors, textures, and glosses. Limit visual confusion.	1,4,5
(06) Compatibility/Contrast	R: Provide appropriate consistency or variety of visual environment. C: Design appearance of elements in a pleasing and harmonious combination.	1,4,5
32 Acoustical Properties		
(01) Sound Generation	R: Control undesirable sound and vibration generation. C: Limit sound generation. Provide specified decibel rating.	2,3,4,5
(02) Sound Transmission	R: Control transmission of sound. C: Design for specified sound transmission classification. Provide STC or SPP rating.	1,2,3,4,5
(03) Reflectance	R: Control reflection, reverberation, and echo production. C: Design for specified reverberation time, and sound path length.	1,2,3,4
33 Illumination		
(01) Level	R: Control quantity of illumination provided. C: Design for specified illumination intensity level. Design to provide specified level of natural light.	2,5
(02) Color	R: Control color (wavelength) of illumination. C: Require lamp color and specified range of correlated color temperature.	
(03) Shadow/Glare	R: Control illumination uniformity. C: Design for specified variation in illumination level over room area.	2,3,4
(04) Reflection	R: Control undesirable reflection. C: Limit reflected light.	2,5
34 Ventilation		
(01) Air Quality	R: Control air quality. C: Design for specified natural ventilation. Design to control rate of air removal and supply design to control odors.	1,2
(02) Velocity	R: Control air movement. C: Design to maintain air motion between specified limits.	1,2
(03) Distribution	R: Control temperature gradients. C: Design to control temperature gradients within specified limits.	1,2
(04) Pressurization	R: Control pressure differental. C: Design to limit air leakage.	2
(05) Temperature	R: Control air temperature content. C: State exterior design conditions. Design to control rate of change of mean radiant temperature within specified range.	
(06) Moisture	R: Control air moisture content. C: State exterior design conditions. Design to provide specified range of relative humidity.	2
35 Measurable Characteristics		
(01) Levelness	R: Control deviation from identified horizontal. C: Require level installation. Design for case of level installation.	5
(02) Plumbness	R: Control deviation from identified vertical. C: Require plumb installation within specified tolerance. Design for case of plumb installation.	5

Test Codes:
1. Design Drawings 2. Design Calculations 3. Laboratory Certification 4. Prototype Testing 5. Inspection

Attribute	Requirement (R)/Criteria (C)	Test
(03) Dimension/Tolerance	R: Control spatial extent for installation or fit within available space.	
	C: Conform to specified spatial dimensions and tolerances.	5
(04) Volume	R: Control volumetric measure or capacity.	
	C: Conform to specified limits of volume or capacity.	5
(05) Flatness	R: Control planar surface characteristics.	
	C: Limit deviation from flat, smooth, or planar surface.	5
(06) Shape	R: Control surface configuration, contour, or form.	
	C: Conform to specified shape limitations.	5
(07) Weight/Density	R: Control weight or density.	
	C: Conform to specified weight or density limitations.	5
36 Material Properties		
(01) Hardness	R: Control resistance to penetration.	
	C: Limit penetration under specified load.	3
(02) Ductility/Brittleness	R: Control capability to shape by drawing. Control tendency to shatter.	
	C: Limit percentage elongation or percent change in cross section before rupture.	3
(03) Malleability	R: Control capability to shape by hammering.	
	C: Limit choice of materials.	
(04) Resilience	R: Control capability to store energy.	
	C: Limit residual deformation after impact load.	3
(05) Elasticity/Plasticity	R: Control capability to retain original shape when load is removed.	
	C: Limit residual deformation after removal of load.	3
(06) Toughness	R: Control capability to change shape without rupture.	
	C: Limit energy absorption before rupture.	3
(07) Viscosity	R: Control fluid resistance to flow.	
	C: Limit coefficient of viscosity.	3
(08) Creep	R: Control permanent change in shape after prolonged exposure to stress or elevated temperature.	
	C: Limit permanent deformation under specified load or temperature conditions.	3
(09) Friction	R: Control tendency of two bodies in contact to resist relative motion.	
	C: Limit coefficient of friction.	3
(10) Thermal Expansion	R: Control change in unit dimension resulting from change in temperature.	
	C: Limit coefficient of thermal expansion.	3
HEADING 4: PRACTICAL		
41 Interface Characteristics		
(01) Fit	R: Control size and shape of interface elements.	1,4,5
	C: Design for physical compatibility with specified elements.	
(02) Attachment	R: Control physical and electrical connection at interface.	1,4,5
	C: Design to use specified connections.	
(03) Tolerance	R: Control variation in interface dimension.	1,4,5
	C: Design to accommodate specified tolerance.	
(04) Modularity	R: Control standardized unit dimensions or repating dimension.	1,4,5
	C: Design for compatibility with the specified module.	
(05) Rotability	R: Control orientation at interface.	1,4,5
	C: Design to provide or permit specified orientations.	
(06) Relocatability	R: Control ability to disassemble move or relocate.	1,4,5
	C: Design to provide specified flexibility to dismount and re-erect.	
(07) Erection Sequence	R: Control order of erection or installation.	1,4,5
	C: Design to provide specified flexibility to dismount or re-erect.	
42 Service		
(01) Repairability	R: Provide for repair or replacement of damaged or in-operative elements.	
	C: Design for ease of repair. Limit use of special tools, limit amount of labor required.	1,4,5

Test Codes:
1. Design Drawings 2. Design Calculations 3. Laboratory Certification 4. Prototype Testing 5. Inspection

(*Continued*)

Attribute	*Requirement (R)/Criteria (C)*	*Test*
(02) Interchangeability	R: Provide for interchangeability of elements. C: Design for interchangeability.	1,4,5
(03) Accessibility	R: Provide access for service and maintenance. C: Design with access panels. Avoid placing connections in inaccessible locations	1,4,5
(04) Replaceability	R: Provide for substitution of equivalent elements. C: Design to permit substitution.	1,4,5
(05) Inconvenience	R: Limit disturbance during maintenance and repair. C: Design to minimize inconvenience. Provide backup or alternate elements.	1,4,5
(06) Extendibility	R: Provide for capability to increase capacity. C: Design to permit or accommodate extension or expansion.	1,4,5
(07) Adaptability	R: Provide for alteration or modification. C: Design to use industry standard connectors and interface elements.	1,4,5
(08) Replacement Sequence	R: Provide for identified order for removal and replacement. C: Design for identified replacement sequence.	1,4,5
(09) Service Frequency	R: Control repair and maintenance frequency. C: Design for identified failure rates and maintenance schedules.	1,4,5
43 Personnel Needs		
(01) Maintenance Personnel	R: Control skill levels required for maintenance. C: Design for maintenance by personnel with identified skills.	2
(02) Training	R: Control availability of trained personnel. C: Require provision for training operators and maintenance personnel.	2

Test Codes:
1. Design Drawings 2. Design Calculations 3. Laboratory Certification 4. Prototype Testing 5. Inspection

Figure 12.5
Suggested Attributes/Requirements/Criteria/Tests

Chapter 13
Specifying for Purchase of Goods

13.1 Introduction to Specifying for the Purchase of Goods

Purchasing is the direct acquisition of materials and equipment by an owner for the owner's use or for installation in the owner's project. The term *goods* is used in purchasing specifications to designate the materials or equipment to be purchased. This term is also used in the Uniform Commercial Code (UCC), which governs most purchasing contracts, so terminology is consistent between the specifications and applicable law.

The goods, which may range from office supplies to furniture, furnishings, and equipment, are purchased from a manufacturer, fabricator, dealer, supplier, or other vendor. The purchasing of goods necessitates a contract or agreement, so the vendor is called the *goods contractor.*

Unlike the construction contractor, the goods contractor is involved in little or no onsite activity such as construction, installation, or erection. This is one of the distinguishing factors of a purchasing contract. The owner may purchase goods under a purchasing contract for installation by the construction contractor. But if the goods contractor is required to participate in the actual construction, the goods would be furnished and installed under a subcontract agreement between the goods contractor and the construction contractor, rather than under a purchasing contract. Equipment, furniture, or other items requiring only assembly and placement, but that are not built into the construction, may still be purchased under a separate purchasing contract. The goods contractor may also perform special services such as providing technical assistance during installation of the goods by others, testing of the goods after installation, start-up procedures, and training of the owner's personnel.

The purchased goods may be for the owner's use, such as for maintenance or a manufacturing process. The goods may also be incorporated into the owner's construction project by another contractor. Figure 13.1 illustrates the case in which delivery and disposition of the goods is made to the construction contractor for installation in the work. In the illustration, a single architect/engineer (A/E) has been contracted by the owner to produce both the purchasing and the construction documents. Purchasing documents that consist of a project manual and drawings should be prepared by an A/E in a manner similar to the preparation of construction documents.

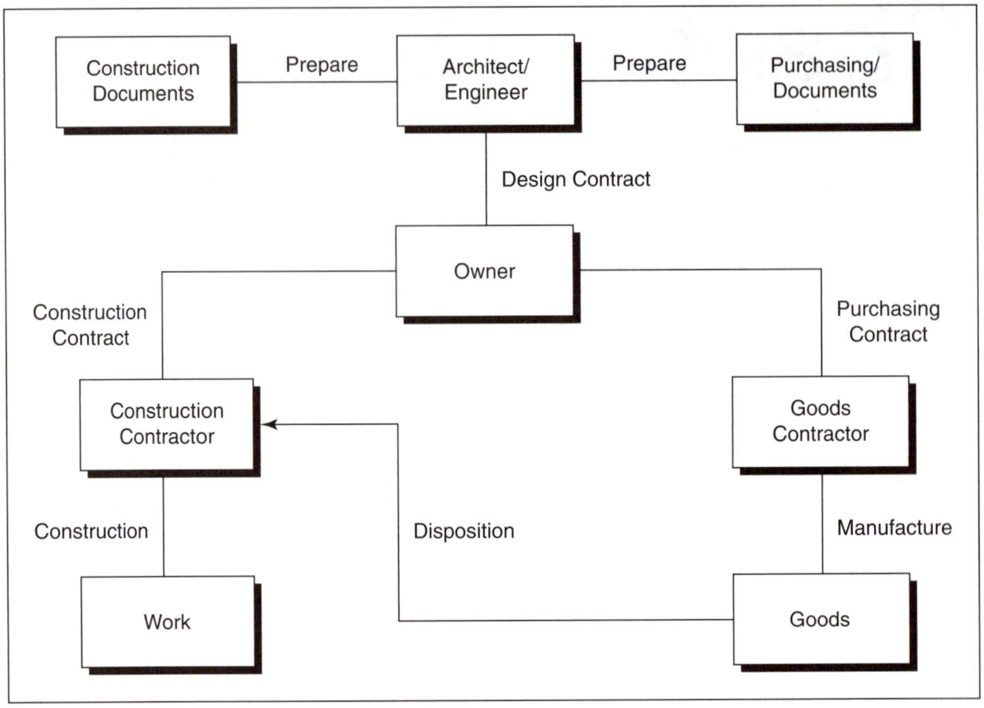

Figure 13.1
Purchasing Contract

13.2 Use of Purchasing

Purchasing documents may be prepared to cover any of a number of activities by the goods contractor:

- To manufacture, fabricate, or supply and transfer to the owner or the owner's representative at the goods contractor's plant
- To manufacture, fabricate, or supply and deliver to a designated location
- To perform additional special services, which may include the preparation of shop drawings, the compilation of operation and maintenance data, technical assistance or guidance during on-site construction or installation of the goods by others, field checking or testing of the goods after installation or erection by others, start-up procedures, and training of the owner's personnel

Each of these activities has particular requirements that must be carefully considered and coordinated throughout the project manual. Note that these activities all stop short of actual on-site construction or installation by the goods contractor. Properly employed purchasing procedures are of appreciable benefit. The uses of purchasing documents to purchase goods in advance of construction or installation include the following:

- To aid in establishing a project budget, as costs of the purchased goods will be known in advance
- To gain lead time on goods that require long manufacture or fabrication periods, and thereby avoid construction delays
- To anticipate shortages of either materials or manufacturing capability
- To take advantage of favorable present cost or to protect against inflation

- To provide information for the design of facilities required to accommodate the specific goods purchased
- To permit acquisition of goods to fulfill the cash allowances of a construction contract by assignment of the purchasing contract to the construction contractor
- To obtain uniformity of function, appearance, or operation and for ease of maintenance, repair, and replacement
- To provide for proprietary or single-source acquisition of goods
- To permit acquisition of goods for which an owner may be entitled to tax exemptions not available to a construction contractor
- To permit acquisition of goods in situations where a construction contractor would not have the necessary technical expertise or the necessary financial or administrative resources
- To reduce the amount of a construction contract in order to reduce bonding requirements or to enable small construction contractors to bid
- To permit the A/E to procure the goods in conjunction with a design construct contract, or the purchasing agency to acquire the goods for an industrial/process engineering organization
- To serve as an adjunct to fast-track construction
- To establish an inventory of goods for later use or installation by the owner

Conversely, acquisition of goods through purchasing contracts may not always be advantageous. The following precautions should be considered:

- The financial investment on the part of the owner prior to use of the goods may be significant.
- The goods may become obsolete before use because of the subsequent production of improved goods.
- The owner and the A/E may invest appreciable administrative time that normally would be provided by the construction contractor if the goods were provided as part of a construction contract.
- A burden may be imposed on the construction contractor as a result of purchased goods being stored on the site.
- The owner may become responsible for time of delivery, handling and storage, maintenance and condition at time of turnover, suitability of items furnished, and completeness of items specified. The owner may therefore be liable to claims by the construction contractor for late or incomplete delivery. This should be considered in establishing liquidated damages in the purchasing contract.
- A particular type and extensive level of expertise is required to develop the purchasing documents.
- The A/E assumes additional responsibilities for coordination of the purchasing and construction.
- The owner or A/E must prepare documents and negotiate and administer more than one contract.
- The construction contractor may object to installing goods that have been purchased separately and may not provide the same warranty as for the items supplied under the construction contract.
- Problems may arise over the assignment of responsibility for performance failures because the goods were acquired under one contract and installed under another.

- Provisions for equipment warranties become more complex and must be carefully addressed to suit specific project needs.
- Certain proprietary goods may be available only from specific dealers and could limit competition and pricing.

13.3 Differences between Purchasing Contracts and Construction Contracts

One of the principles of purchase specifying is that on-site activities by the goods contractor are limited, and no construction or installation is usually involved. The solicitation, procurement instructions, bid or proposal forms, agreement, conditions of the contract, and Division 01—General Requirements used in purchasing contracts will differ from those used in nonpurchasing construction documents. Most of these differences are readily apparent. The more subtle differences occur in the conditions of the contract. Several subjects should be carefully considered:

Definitions The definitions commonly found in construction documents, particularly those related to on-site construction or installation activities, are not always applicable to purchasing documents. Inapplicable definitions must be avoided. It is possible for both a purchasing contract and a construction contract, which includes installation of the purchased goods, to be applicable at the same time. Although definitions of the same word or term may appear in the documents for both contracts, there may, of necessity, be a slight difference between them. It is imperative that the various definitions not conflict with one another.

Bonds Depending on the type of goods to be purchased and their intended disposition, a performance bond should usually be required. The bond provides recourse for the owner in the event of nonperformance in the manufacture, supply, or delivery of goods.

Insurance The types of insurance needed for purchasing contracts depend entirely on the specific activities required of the goods contractor. Manufacture or fabrication will generally be at the goods contractor's plant, and so will not usually be subject to direct control or supervision by the owner. Liability insurance coverage should be part of the goods contractor's normal business operation. Stipulations are therefore generally inappropriate, except that property insurance should be required to protect the owner's interest in undelivered goods for which payment has been made and for goods that have been delayed in shipment at the owner's request.

When the purchasing contract includes delivery by the goods contractor to a designated location, the documents should require the goods contractor to provide transit insurance to cover the goods during shipment, especially if any payment has been made. Receipt by the owner or construction contractor at the point of delivery will generally involve the transfer of responsibility. Insurance after delivery is usually a responsibility of the owner.

Technical assistance during onsite construction or installation, field checking or testing after installation, start-up procedures, or training of the owner's personnel will all directly involve the goods contractor in on-site activities. Although the goods contractor's exposure to insurable risk under the purchasing contract may be limited in comparison to a construction contract, the exposure still exists. The documents for such purchasing contracts should require the goods contractor to provide liability insurance during on-site activities similar to the insurance provided under construction contracts. Coverage should include workers' compensation, general liability, and automobile liability coverage. All decisions concerning types and amounts of required insurance should be made by the owner in consultation with legal counsel and insurance advisers.

Goods Contractor's Responsibilities Generally, the goods contractor should not be charged with responsibilities directly related to on-site construction. The goods contractor should not be required to visit the site or conduct on-site investigations unless field measurements are required or existing field conditions affect production of the goods, or unless technical assistance is specified. Matters such as construction permits, licenses, construction equipment, temporary facilities, and safety should be the responsibility of the construction contractor.

Unless the goods to be purchased are exceptionally expensive or complex, subcontractors are usually not involved in a purchasing contract. Stipulations concerning subcontractors, including relationships between contractor and subcontractors, or owner and subcontractor, are therefore not included in most purchasing documents.

Except when unique codes or laws govern, as in the case of the purchasing of nuclear components, pressure vessels, or fire protection systems, it should not be necessary to stipulate the goods contractor's compliance with laws and regulations (other than applicable standards such as the American Society for Testing and Materials [ASTM], American National Standards Institute [ANSI], American Society of Mechanical Engineers [ASME], or National Electrical Manufacturers Association [NEMA]) pertaining to manufacturing processes or to specify provisions for safety and protection. Manufacture or fabrication remains solely the goods contractor's responsibility as part of normal business operations.

A/E Responsibilities If the A/E is involved in preparing construction documents under which the goods are to be installed, then the A/E must coordinate the purchasing provisions with the construction contract. Proper relationships must exist between the contracts, and suitable provisions must be stipulated in each so that pertinent items are neither omitted nor duplicated. Receipt and handling of the goods, utility connections, start-up procedures, training of the owner's personnel, and similar items must be clearly stipulated.

The responsibilities of the A/E for a purchasing contract vary according to project requirements and the owner's desires. Responsibilities must be carefully established and thoroughly understood by both the owner and the A/E. The A/E may or may not be obligated to visit the goods contractor's plant or to conduct or witness shop tests. The A/E may or may not become involved in observation or verification of the goods at time of delivery. These items will materially affect the provisions of the purchasing contract as well as the A/E agreement for professional services, and must be clarified with the owner in advance.

Shop Drawings, Product Data, and Samples Review and approval of submittals are normally managed by the A/E. Stipulations covering submittal of shop drawings and product data will be similar to the requirements generally found in construction contracts. A greater-than-usual number of approved submittals may be acquired for records and for transmittal to the construction contractor or to other end users of the purchased goods. It may also be advantageous to make this available during solicitation on the construction contract.

Operation and Maintenance Manuals Documents for traditional construction contracts frequently include a requirement for operation and maintenance manuals, particularly for complex equipment or systems. These manuals may be even more important under a purchasing contract. There may be little direct contact between the goods contractor and the eventual installer or erector; therefore, the information supplied by operation and maintenance manuals may be of critical importance.

Payment, Completion, and Acceptance Payment procedures depend largely on the nature of the goods being purchased. Items may be such that partial deliveries and corresponding progress payments will be made. For some complex equipment, progress payments may sometimes be made during manufacture or fabrication. In most instances, however, procedures will be based on making a payment at each of three stages:

- Approval of submittals
- Delivery to the designated location
- Final acceptance

If progress payments are to be made while the goods are being manufactured and remain under the control of the goods contractor, provisions should be included for protecting the owner's interest in the partially completed goods. Appropriate contract language for this purpose should be obtained from the owner's legal counsel.

As the A/E may be involved in observation or verification of the goods delivered, the corresponding obligations should be considered in establishing procedures for acceptance and payment. The A/E should not assume unnecessary responsibilities nor risk undue exposure on matters of which it has no direct knowledge or control.

Procedures for final acceptance and final payment also depend on the nature of the goods. Some items are complete and fully acceptable at the time of delivery, and final payment is made accordingly. For other items such as equipment, final acceptance is reserved until the item has been installed or placed in service and proven acceptable. However, in fairness to the goods contractor, some reasonable time limit should be set between the time of delivery and the time of installation, after which final payment should be made regardless of whether the item has been installed and tested. In any case, the procedures to be followed must be clearly specified.

Although goods delivered to the site are usually covered by property insurance, measures may need to be taken to prevent theft or damage.

13.4 Uniform Commercial Code

The UCC governs contracts dealing with the sale of goods and, with minor variations, is the law in every state except Louisiana. The purpose of the UCC is to facilitate commerce by providing certainty and consistency in commercial transactions. Inapplicable provisions may be waived by mutual agreement of the parties to the contract; otherwise, the UCC applies to every situation falling within its terms. The A/E should be aware of the basic provisions of the UCC, which become effective in the absence of, or may sometimes actually override, written contract provisions. The UCC is published by each of the using states as a part of its state statutes.

The UCC is composed of articles that govern a wide variety of frequently recurring business transactions, including the sale of goods, negotiable instruments, and financial transactions involving security interests. Article 2, dealing with sales, has the greatest effect on purchasing contracts. Its coverage is limited to sales of movable goods, so contracts that are primarily for services are excluded. The term *services* includes activities by the goods contractor such as technical assistance or an installation check.

The following four basic areas of contract law are covered by Article 2 of the UCC.

13.4.1 Contract Enforceability and Formation

Contracts for the sale of goods costing more than $500 must be in writing to be enforceable. Generally, a written contract without ambiguities will be construed as the total agreement between the parties and may not be contradicted by a prior agreement or a concurrent oral agreement. The strict common-law contract rules relating to offer and acceptance and to the formation of a contract are somewhat relaxed under the UCC. The thrust of Article 2 is to establish a contract if at all possible, even though there was never a true meeting of the minds, in the contract-law sense, as to all of the various terms and conditions. The legal test is whether the parties intended to make a contract. Terms, including price, may be left open, to be determined at a later date.

13.4.2 Warranties

The UCC attaches certain warranties, expressed and implied, to sales transactions and govern warranties provided by the seller. Generally, a seller warrants the transfer of clear title, free of any security interest of which the buyer is unaware.

Express warranties may consist of:

- An affirmation or promise, such as a claim of performance
- A description of the goods such as photos, drawings, or specifications
- A sample or model that becomes a basis of the agreement

There are also two implied warranties under the UCC:

- *Merchantability.* If the seller is a merchant who deals with goods of the kind being conveyed, a warranty is implied that the goods are fit for the ordinary purposes for which such goods are used.
- *Fitness for a Particular Purpose.* If the seller has reason to know of any particular purpose for which the goods are required and that the buyer is relying on the seller's skill or judgment to furnish suitable goods for that purpose, an implied warranty of fitness for that particular purpose arises.

13.4.3 Contract Interpretation

The UCC acknowledges the common commercial practice by which the contracting parties leave certain terms to be agreed upon at a later date, or to be based on reasonable commercial practices. Quantity must generally be agreed upon; however, details of payment, delivery, and even price may be supplied by the UCC if they are not covered by the written contract. Unless otherwise agreed, the UCC allocates risk of loss or damage to the goods between the buyer and the seller, depending on the extent to which the agreement has been performed. FOB (free on board), CIF (costs, insurance, and freight), and similar shorthand mercantile terms that place responsibility for transportation costs, insurance, and risk of loss are defined in the UCC.

The parties to an agreement governed by the UCC are bound by a requirement that they act in good faith, and unfair or unconscionable contracts are not enforceable.

13.4.4 Remedies

The UCC provides remedies for the seller in the event that the buyer becomes insolvent or breaches the contract by wrongfully rejecting the goods, refusing to pay, or revoking acceptance. The seller may elect to stop delivery, reclaim the goods, recover the price, or recover damages. Conversely, if the seller is at fault for failing to deliver the goods as specified, the buyer may cancel the contract, purchase a substitute, and recover damages. The UCC provides alternative bases for computing actual damages and allows the recovery of incidental and consequential damages if appropriate. A liquidated damages clause is enforceable if the amount is reasonable and verifiable, but if the amount is excessive and unsubstantiated, the clause is unenforceable.

13.5 Format

Assuming the goods are to be competitively bid, *MasterFormat®*, *SectionFormat™*, and *PageFormat™*, as well as other Construction Specifications Institute (CSI) organizational tools, may be used to great advantage. The organization, concepts, locations of subject matter, and language established by CSI can be adapted to purchasing documents. Use of documents similar to those already accepted and familiar to goods contractors and suppliers in the construction industry helps avoid confusion. Usually, a purchasing contract for goods and a construction contract that includes installation of those goods will be in force at the same time, in which case the two project manuals should be similar and must be carefully coordinated.

13.5.1 Procurement and Contracting Requirements

A purchasing project manual customarily contains procurement requirements, contracting forms, conditions of the contract, and specifications. These documents are similar to those used for construction projects, but reflect the differences previously discussed.

The procurement requirements might include prebid information, instructions to bidders, information available to bidders, bid forms, and bid security forms. An agreement form, performance bond, general conditions, and supplementary conditions should also be elements of the purchasing project manual. Standard documents available for purchase projects are discussed at the end of this chapter.

13.5.2 Division 01—General Requirements

Division 01 specifications are as important for purchase projects as they are for construction projects. These sections should be used to expand the provisions of the general conditions and to provide general procedural requirements that will govern the other specification sections. The Level 2 and Level 3 sections listed in *MasterFormat®* fall into three categories with respect to purchasing contracts.

The first category includes those sections that are generally applicable to all purchasing contracts:

Section 01 10 00—Summary

Section 01 30 00—Administrative Requirements

Section 01 40 00—Quality Requirements

Section 01 60 00—Product Requirements

The second category includes those sections that are sometimes applicable to purchasing contracts, the determining factors being the level of activity required of the goods contractor and the particular aspects of the project:

Section 01 20 00—Price and Payment Procedures

Section 01 70 00—Execution and Closeout Requirements

Section 01 80 00—Performance Requirements

The third category includes those sections that are generally not applicable to purchasing contracts because they relate to on-site activities:

Section 01 50 00—Temporary Facilities and Controls

Section 01 92 00—Facility Operation

Section 01 94 00—Facility Decommissioning

13.5.3 Divisions 02 through 49

Use of the appropriate divisions and sections of the technical specifications will be determined by the project requirements, just as they are in construction contracts. Depending on the goods to be purchased, the project manual may contain one or more specification sections. Many of the *MasterFormat®* titles deal exclusively with construction and would not be appropriate for purchasing, but others pertaining essentially to furnishings, furniture, equipment, or materials may be used. *SectionFormat™* presents a format for organizing a specification section that readily lends itself to the specific requirements of a purchase project.

Of the three parts of the *SectionFormat™* (PART 1—GENERAL, PART 2—PRODUCTS, and PART 3—EXECUTION), the third part, which deals with on-site activities, will be the least used in purchasing. Items related to the disposition of goods and special services required of the goods contractor would typically be stated in PART 3. However, if there are no requirements to be stated in PART 3, one can simply note "Not Used." Development of individual articles under each of the three parts will be determined by the basic purchasing procedures; the nature of the goods; whether the section is written as a proprietary, descriptive reference standard, or performance specification; the required level of contractor activity; and the eventual disposition of the purchased goods.

13.5.4 Disposition of Goods

Preparation of purchasing documents must include consideration of the eventual disposition of the purchased goods, as each of the methods of disposition will have its own effect on various document sections. Three basic methods should be considered:

1. *Retention by Owner for Own Use and Installation.* This is the simplest method of disposition, with the owner establishing a stockpile or inventory of goods for future use. The owner purchases the goods, with or without delivery by the goods contractor, and thereafter assumes full responsibility for them. After completion and final payment, the A/E and the goods contractor should no longer be involved, except for the goods contractor's warranty responsibility.
2. *Furnished by Owner for Installation by Others.* In this case, the owner is involved in a construction contract and the purchased goods are to be furnished for installation as part of the construction project. The owner is usually involved in receipt, storage, and maintenance of the goods, and possibly in delivery to the construction contractor. The owner is responsible to the construction contractor for suitability, completeness, and condition of the goods at the time of transfer, and the construction contractor thereafter becomes responsible for the storage and handling of the goods and for their installation. The construction contractor includes in the construction contract the cost of storage, handling, and installation of the goods and must therefore be provided with all pertinent information. Transfer of any warranties for the goods to the construction contractor might be doubtful, and the owner, as the original purchaser, may be required to retain and enforce the warranties.
3. *Assignment of Purchasing Contract to Construction Contractor.* Under this arrangement, the construction contractor accepts the purchasing contract, similar to a purchase order, and assumes all responsibilities for receipt, payment, and administration in addition to installation. The construction contractor again must be provided with all pertinent information prior to determining contract cost. The owner may lose a part of any cost advantage that may have been gained through purchasing, but will retain other inherent advantages.

In any case, the method of disposition must be clearly stipulated in the purchasing documents. If a construction contract is also involved, the method of disposition must be stipulated in the construction contract as well, and the procedures definitively established with both the goods contractor and the construction contractor. The documents must state which items, appurtenances, accessories, or other parts are to be furnished under each contract and must establish responsibilities for the remainder. These procedures require particularly close study and coordination by the A/E.

13.6 AIA Purchasing Documents

American Institute of Architects *(AIA) Document A271, General Conditions of the Contract for Furniture, Furnishings and Equipment*, and its companion Instruction Sheet were produced in conjunction with the American Society of Interior Designers. AIA also publishes *Document A171, Standard Form of Agreement Between Owner and Contractor for*

Furniture, Furnishings and Equipment, and its companion Instruction Sheet to accompany AIA Document A271. AIA Documents A271 and A171 reflect a number of the principles set forth in this section, but also differ in some respects. The documents state that they depart in certain matters from *AIA Document A201, General Conditions of the Contract for Construction*, and that the departures include some duties of the architect, responsibilities of the owner, and requirements for payment and completion. The documents include references to stipulations of the UCC. The documents also contain requirements for the goods contractor's inspection of the premises, references to subcontractors, installation requirements, and full insurance requirements. Items inappropriate to a given project can, of course, be deleted by the supplementary conditions. AIA Document A271 is appropriately used for interior design contracts, and many of its features are adaptable to the purchasing of other types of materials and equipment.

13.7 EJCDC Procurement Documents

Whereas CSI and AIA use the term *purchasing* to define goods purchasing, the Engineers Joint Contract Documents Committee (EJCDC) uses the term *procurement.* The following paragraphs are excerpted and paraphrased from EJCDC P-001, Commentary on Procurement Documents, and are updated to reflect the more recent editions of published documents.

These EJCDC documents are used when an owner purchases goods directly from the provider for use in construction projects. These purchasing documents are an alternative to accepting standard purchase order forms provided by manufacturers and suppliers and avoid the inadequacies of such forms, as well as the difficulties of adapting them to properly address the owner's needs and the project's requirements. There are four standard EJCDC purchasing documents:

- *EJCDC P-200, Suggested Instructions to Bidders for Procurement Contracts*
- *EJCDC P-520, Suggested Form of Agreement Between Buyer and Seller for Procurement Contracts*
- *EJCDC P-700, Standard General Conditions for Procurement Contracts*
- *EJCDC P-800, Guide to the Preparation of Procurement Supplementary Conditions*

13.7.1 General Considerations

Purchasing documents are closely integrated, and the terms used in all four standard purchasing documents are defined in the general conditions. A change in any one of these documents may require a change in the others. Anyone using the EJCDC purchasing documents would do well to read those parts of *EJCDC COM-01, Commentary on Agreements for Engineering Services and Construction-Related Documents*, which pertain to the comparable forms for construction contracts. Many of the ideas and suggestions discussed there are not repeated in this brief commentary. No standard purchase bid form has yet been prepared, but *EJCDC P-400, Suggested Bid Form and Commentary for Use*, may be used as a guide when preparing a purchase bid form. In preparing the purchasing documents, EJCDC paid close attention to the suggested locations of subject matter as set forth in *EJCDC 1910–16, Uniform Location of Subject Matter*. In particular, note the frequent references in the supplementary conditions to the proper place for addressing a given topic.

13.7.2 Intended Use

The purchasing documents are not intended for use where the manufacturer or supplier of the goods is required to participate in the installation at the project site. The EJCDC documents assume that either the owner's personnel or a construction contractor or subcontractor at the site will install the furnished goods. When the contractor is required to "furnish, deliver, and install," the project is more like a traditional construction contract, and the standard forms for construction contracts should be used rather than purchasing documents. If the purchasing documents are to be used in cases where there are multiple deliveries of the same item, changes may be required to the supplementary conditions.

13.7.3 Definitions

In the interest of uniformity, many terms used in the EJCDC documents for construction contracts are defined in the purchasing documents as having the same or similar meanings. Accordingly, the manufacturer or vendor furnishing the goods is referred to as the contractor who has entered into an agreement with the owner. A person or entity referred to in the documents as a manufacturer, supplier, fabricator, distributor, or vendor would be a party having a subcontract with one of the prime construction contractors or a sub-subcontract with one of the subcontractors performing work at the project site. The material and equipment to be furnished by the goods contractor are defined as *goods*. Frequently, the goods contractor will be required to be present at the project site during installation, testing, or starting of the goods or to instruct the owner's personnel or provide other consulting or advisory services. Such construction site services are referred to as *special services*. Special services are not intended to cover installation services by the goods contractor because the purchasing documents are not intended to apply to a *furnish, deliver,* and *install* relationship. The various categories of special services to be furnished will be described generally in Article 1 of the agreement and particularized in the supplementary conditions or the specifications. The term *other services* refers to all other services that may be reasonably inferred from the purchasing documents as being necessary to furnish the goods and special services. The term *work,* defined in construction contract documents, is not used in the purchasing documents.

Chapter 14 Federal Agencies

14.1 Specifying for Federal Agencies

Many federal agencies administer construction programs, and each has different construction document preparation requirements. This section identifies some of the unique requirements involved in preparing contract documents for federal agency projects. Many of the principles discussed here will also apply when the client is a city, county, or state government agency, or a school district with specific requirements for the form and content of their construction documents.

MasterFormat® is often the common link between federal government and private-sector specifications. Most government agencies use *MasterFormat®* to organize their guide specifications. However, their section numbers and titles may sometimes differ from those in the current edition. In addition, federal agencies sometimes intentionally deviate from *MasterFormat®* to accommodate their unique requirements.

Most government guide specifications conform to *SectionFormat™*. However, the method of numbering paragraphs may not always follow the preferred alphanumeric arrangement in the suggested *PageFormat™*. Some agencies number their paragraphs and subparagraphs in a legal style, Arabic numeral format (e.g., 1.1, 1.1.1, 1.1.2, 1.1.2.1). Some agencies also require the specification of procedural matters for maintenance information and shop drawings to be included in each technical specification section.

14.2 Procurement and Contracting Requirements

Federal agencies usually do not require the architect/engineer (A/E) to prepare all the components of a document package that will be used for solicitation of bids on a construction contract. In some cases, the A/E will prepare the general requirements and technical specifications, and the agency will prepare the procurement and contracting requirements. The A/E must be familiar with these client agency procedures before preparing the specifications. When agency-prepared documents are used, the content of Division 01 will vary greatly from a Division 01 prepared for use in the private sector.

The first difference regarding the conditions of the contract for a federal project is the name of the document. Unlike the American Institute of Architects (*AIA Document A201,*

General Conditions of the Contract for Construction) or the Engineers Joint Contract Documents Committee (*EJCDC C-700, Standard General Conditions of the Construction Contract*), the federal documents might be titled "General Provisions," "General Standards," "Special Provisions," or "Contract Clauses." They are all based on the Federal Acquisition Regulations (FAR).

There are other significant ways in which federal general conditions differ from their private-sector counterparts:

- A contracting officer, rather than the A/E, is responsible for administering the procurement and construction phases of the contract. The contracting officer is either a military officer or a civilian employee of the government. Consequently, the A/E does not exercise the same authority on federal projects that it ordinarily would on projects in the private sector.
- The federal government is both the owner and the primary regulating authority.
- Local building code authorities do not have jurisdiction on federal property, although other federal agencies may have authority. Local authorities having jurisdiction (AHJs) do, however, have jurisdiction on private property leased by federal agencies. Federal guide specifications may include requirements that take exception to or exceed local building codes.
- The federal government is self-insuring and does not procure commercial insurance for its buildings or their contents. Federal guide specifications may contain more stringent requirements for fire protection, safety, security, and component and construction durability than those in the private sector.
- The means of settling contract disputes is different from the means encountered in the private sector. Although this normally does not directly affect the preparation of specifications, it does have a significant impact on administration of the contract.

14.3 Social Value Regulations

Federal government contracts may include certain "social value" regulations designed to assist particular industries and other special interest groups. Some examples are wage rates, environmental regulations, preference for American shipping, affirmative action, utilization of small businesses, utilization of businesses owned by women and minorities, labor standards, environmental issues, and accessibility standards.

The Buy American Act is another such provision. It gives U.S. manufacturers an advantage by imposing restrictions on the use of foreign-made products. A product is considered to be a U.S.-manufactured product if it is manufactured in the United States and at least 50 percent of its components, by cost, are mined, produced, or manufactured in the United States. Some materials are exempt from Buy American provisions, such as raw materials of which there are limited domestic supplies. Use of most other foreign products is generally prohibited unless the contractor can furnish reliable evidence that the cost of an equivalent U.S.-made product will be more than 6 percent greater than the cost of the foreign-made equivalent, or if the foreign product is used in a unique situation where no domestic product can meet the criteria.

14.4 Use of Manufacturers' Names

In almost all situations, federal agencies prohibit the use of brand names, manufacturers' names, or other proprietary or restrictive requirements for products. The intention is to encourage unlimited competition consistent with the type and requirements of the work and to maximize the use of standard products and current models meeting the functional requirements of the project. Specifications for acquisitions should state only the government's actual minimum requirements and describe the supplies or services in a manner that will encourage maximum practicable competition. Brand names may be used in federal government specifications only under the following exceptional conditions:

- When there is no other feasible way of describing the essential functional or physical characteristics (such as to describe the color of a paint, appearance of stone, or the performance of a high-technology product).
- When an agency is extending or connecting to an existing system and the new portion or components must match the existing system. In this situation, the agency would prepare a justification and approval document for procurement of only one brand. Where a specific brand name is required, the A/E may be required to provide technical input for preparation of justification and approval as part of the specification preparation effort.
- When specifications include a statement explaining that the brand names are included only to describe stated salient characteristics such as color, pattern, and operational characteristics, and that other products having the same characteristics will not be excluded.
- When specifications list a minimum of three products and state specifically that acceptable products are not limited to the manufacturers of those products.

When brand-name descriptions are necessary, the specifications must clearly identify and describe the particular physical, functional, or other characteristics of the brand-name items that are considered essential to satisfying the requirement.

In order to eliminate any chance that the use of brand names in specifications might limit competition, the brand-name or equal clause should be included by the contracting authority for the agency whenever brand names are used, as previously described.

In some cases it is necessary to limit the contractor to the use of a single product for a particular function and identify that product by brand name and model number. These proprietary specifications are the exception to the generic specifications typically used in federal specifications, and should be used only when no other product can accomplish the given task. The A/E is required to provide a technical justification to support this proprietary requirement. The agency will review the justification and approve or disapprove the inclusion of the proprietary specification in the project specifications.

14.5 Reference Standards

There are four classifications of reference standards used for specifying government construction. Listed in the order of preferred usage, they are private-sector standards, commercial item descriptions (CIDs), federal specifications, and military specifications.

Federal specifications were originally established to meet a vital need, and many of them also have set standards for industry. However, current government policy encourages development and use of nongovernment standards to replace military specifications and federal specifications. Private-sector organizations such as the ASTM International (ASTM), the National Fire Protection Association (NFPA), and American National Standards Institute (ANSI) are now more active in developing standards for construction products, and federal specifications are being replaced by these standards, or by CIDs, which include references to ASTM, ANSI, and other industry standards. Government policy currently favors the use of such private-sector standards when they exist and when they meet the needs of the agency concerned. It is current U.S. Department of Defense (DOD) policy to convert military specifications to private-sector standards or federal specifications when used in nonmilitary construction. Many federal specifications have not been updated and, providing the owner has no objections, should not be used when more current standards are available.

Federal specifications are documents issued by the General Services Administration (GSA) Federal Supply Service and cover a wide variety of items used by the federal government, including some construction products. Federal specifications generally appear as a reference under PART 2—PRODUCTS of a specification section. Military specifications are similar, except that they are issued by the DOD and include only a few construction products. Military specifications are organized with various classifications of type and quality.

Most federal specifications are classification documents that include requirement levels of the product described and grade the product in accordance with its physical and functional characteristics. When referencing a federal specification, it is usually necessary to state the particular class or grade of the product being specified.

14.6 Specification Items for Consideration during A/E Contract Fee Negotiation

Preparing a specification for a federal agency may take more time than a specification for a private-sector owner, including the time required to read agency manuals and other documents containing requirements relating to the project. Consider the following items when the contract fee is being negotiated:

- What is the form of the available master guide specifications (e.g., printed copy, electronic files that might require special software or text files that might not be compatible with the A/E's computer or software)?
- Who develops or provides specification sections that are not included in the agency-furnished specifications?
- How many reviews are required by the client agency?
- In what form are the specifications to be presented for each review?
- How many copies are required for submittal at each review?
- Will the agency provide the front-end documents, or will the A/E be required to write them?

- What effect will this have on preparation of the remainder of the contract documents?
- What is the time frame for reviewing submittals and for review by the agency?
- Will the agency's own page format require additional work?
- Will the agency require all sections to be produced in the same typeface (e.g., a mixture of printing developed by separate disciplines may not be accepted)?
- Will reference documents need to be acquired and, if so, will the client agency provide copies or reimburse their acquisition cost?

14.7 Federal Government Agencies' Construction Documents Policies

Each agency involved in construction has its own policies and procedures for administering construction programs. Each agency also publishes its own construction documents and forms or has combined with other federal agencies to publish common documents. In Divisions 01 through 49, federal guide specifications are tailored to the specific project by the A/E. Procurement requirements and contracting requirements are later incorporated into the contract documents by the agency before issuance of the project for procurement. Figure 14.1 illustrates policies followed by several federal agencies in issuing documents to A/E firms for use in construction contracts.

14.8 Federal Government Guide Specification Program

The GSA and the National Institutes of Health have adopted a commercially available master guide specification for use on their projects. The GSA uses the complete specification system with some exceptions. The master guide specifications are coordinated with *MasterFormat*®. A/Es working on a GSA project should obtain the appropriate front-end documents. Most other federal agencies involved in construction have developed master guide specifications independently to meet their own particular needs. Various groups and individuals have recommended uniform specifications development to minimize duplication of effort and to make it easier and less costly for contractors, A/Es, manufacturers, and others that do business with the federal government. Since the advent of the National Institute of Building Sciences (NIBS) and the Construction Criteria Base (CCB), most government agency specifications are available on CD, DVD, or the Internet. SpecsIntact is a free download.

14.8.1 Construction Criteria Base

The CCB grew out of a study of problems experienced in disseminating federal master guide specifications for construction.

The system was developed for electronic dissemination of information such as master guide specifications, technical manuals, standards, and other documents. Its objective is to

Agency Name	Procurement and Contracting Requirements		Divisions 01—49			
	Agency Supplies	Agency Supplements	A/E Supplies	Agency Supplies Guide Specifications	Agency Approves A/E Specifications	Agency Supplements A/E Specifications
COE	x			x		
EPA		x				x
FmHA	x				x	
FAA		x			x	
FBOP	x		x	x		
FHWA			x			x
FHA	x			x		
FS	x			x		
GSA	x					
HUD		x			x	
NASA	x			x		
NAVFAC	x			x		
NIH		x			x	
NPS	x				x	
SCS	x				x	x
USPS	x		x	x		
VA	x			x		

Acronyms for the definition of abbreviations used in this figure.

COE	Department of the Army, Corps of Engineers	NASA	National Aeronautics and Space Administration
EPA	Environmental Protection Agency		
FmHA	Farmers Home Administration	NAVFAC	Naval Facilities Engineering Command
FAA	Federal Aviation Administration	NIH	National Institutes of Health
FBOP	Federal Bureau of Prisons	NPS	National Park Service (agency of the Dept. of Interior)
FHWA	Federal Highway Administration		
FHA	Federal Housing Authority	SCS	Soil Conservation Service
FS	Forest Service	USPS	United States Postal Service
GSA	General Services Administration	VA	Veterans Administration
HUD	Department of Housing and Urban Development		

* GSA uses a commercially available master guide specification system.

Figure 14.1
This Chart Illustrates the Policies Followed by Several Federal Agencies in Issuing Documents to A/E Firms for Use in Construction Contracts

put information at the user's fingertips as quickly and cost-effectively as possible. Use of the CCB is intended to improve the quality of construction, decrease overlap and conflicts that exist among criteria, speed introduction of new technology into the construction process, and improve dissemination of master guide specifications and other design and construction documents. Information on the CCB is updated semiannually and is easily transferred to personal computers.

The CCB includes the combined Unified Facilities Guide Specifications (UFGS) from the DOD, including the Army Corps of Engineers (COE), the Naval Facilities Engineering Command (NAVFAC), and the Air Force. Also included are specifications and documents from the Veterans Administration (VA), Federal Aviation Administration

(FAA), Federal Highway Administration (FHWA), GSA, and National Aeronautics and Space Administration (NASA). The system includes specifications processing software called SpecsIntact. The CCB CD/DVD also includes handicap accessibility regulations, asbestos regulations, FAR, Occupational Safety and Health Administration (OSHA) standards, federal standards, and some private-sector standards.

SpecsIntact, meaning "specifications kept intact," is an automated system created by NASA for developing construction project specifications. UFGS and NASA specifications are formatted for editing in the SpecsIntact system. SpecsIntact features include verification that references listed in the front of each section are actually cited in the document, submittal verification, generation of submittal lists, paragraph renumbering, note deletion, testing requirement report generation, page numbering options, creation of tables of contents, section reference verification, use of multiple specification sources, and online help. Once the appropriate specifications have been located and listed, the user can access SpecsIntact, which will process a specification section tailored to the requirements of the project. SpecsIntact embeds notes into the body of the specification section. The software includes a feature that will delete the notes during the print process; however, the notes stay in the working file until manually deleted.

Development, improvement, and enhancement of the CCB continue. Some federal agencies may insist that the CCB be used to prepare contract documents for their projects. When this is the case, the subscription cost and necessary equipment should be considered in fee negotiations.

14.8.2 Obtaining Master Guide Specifications

If only hard-copy master guide specifications are available from the client agency, the A/E's request for copies should be made in the early phases of project development, and selections should be made from the most current list of section titles available from the client agency. The request for copies of master guide specifications should take place after the agency's review and approval of the preliminary design. If requested by the client agency, this first submittal could include outline specifications that were not derived from the agency master guide specifications or only a listing of proposed specification section titles. Such a listing can serve as a basis for the request for copies of master guide specification sections. As the project design develops, it is often necessary to request additional sections.

The methods of acquiring specifications from government agencies are varied. In the past, the only option was to obtain a hard copy of the agency specification and have it completely retyped. This is very time consuming and should be considered during fee negotiation if this is the process to be used.

14.9 Using Federal Government Master Guide Specifications

Before entering into a design contract with a federal agency, the A/E should determine whether the agency has master guide specifications, whether their use is required, and whether copies will be provided free of charge. The time required to produce project specifications from agency guide specifications is subject to many variables. Chief among these is the A/E's degree of familiarity with the required methods of specifying products and systems within the client agency guide specifications (e.g., extensive use of reference standards and general descriptions in lieu of manufacturers' names and model numbers). The greater the differences between the client agency documents and the A/E's office master guide specification, the more time will be required. Difficulties may be compounded if the client agency master guide specifications do not conform to CSI principles and practices, and if the section numbers and titles differ considerably from the current edition of *MasterFormat*®. If differences exist between the agency's specifications and *MasterFormat*®, the A/E is usually expected to use the numbers and titles defined by the agency.

14.9.1 Editing Master Guide Specifications

The agency master guide specifications are just that—guides—and must be edited to suit each project. Guide specifications are usually written for new construction and may require considerable modification for repair or rehabilitation projects. Where coverage is lacking, the A/E must provide the needed material in an agency-approved form. Individual guide specification sections may often be much more detailed than those to which the A/E is accustomed. For example, some guide specification sections might be over 100 pages long. If help is needed to interpret master guide specifications, most agencies will provide assistance. A number of guidelines should be followed when editing agency master guide specifications:

- The sections should be tailored to suit the project, not simply copied in their entirety.
- Identification markings should be deleted on the master (such as headers or footers with the section number or revision number, and revision notations within the text of the section). If the UFGS or NASA specifications are used, the SpecsIntact software is capable of doing this electronically.
- The sections should be edited to avoid conflicts related to products occurring in more than one section.
- Additional information should be added as necessary.
- Paragraphs should not be selected by title alone.
- The entire section, not just the bracketed text, should be read and edited.
- All required selections should be made and all blanks edited.
- Applicable requirements from the editing notes should be incorporated.
- Requirements in the editing notes should be coordinated by the A/E. If SpecsIntact is used, the software is capable of deleting editing notes electronically.
- The references listed at the beginning of the section should be edited to include only the standards remaining within the body of the edited specification. If Specs Intact is used, the software is capable of doing this electronically.
- Each reference listed should indicate the current issue or revision date, unless there is a specific reason to reference an older standard or unless the agency directs otherwise.
- The written instructions from the client agency should be carefully followed for page format, layout, and placement of the agency's contract number. Most agencies are very explicit concerning these details.

- All specifications should be proofread; do not depend on client agency reviewers to find errors and discrepancies.
- The index in the front of each section should be deleted after the editing process is complete, unless the agency requires otherwise. If SpecsIntact is used, the software is capable of producing an index electronically.
- If the agency requires submittal of the marked-up section:
  - A readable copy of the original should be used so that it can be reproduced for the submittal.
  - Use of a highlighter on a review copy should be avoided, as some highlighting photocopies solid black.
  - Pencil should be used for additions and deletions, marked dark enough so notations will photocopy clearly. If permitted by the agency, make the insertions and deletions electronically using the redline and strike-out features of the word-processing software.
  - Some agencies require redlining on the electronic media rather than markups of hard-copy originals. Agency requirements should be followed in this regard.

If the client agency does not have a guide specification for a required section, the A/E must prepare the section using other resources. The new section must conform to the agency standards for format and numbering and must avoid proprietary or restrictive product requirements. The specifications should permit unlimited competition consistent with the type and quality of work. They should be prepared based on standard products and current models of equipment that meet the functional requirements of the facility.

14.10 The Review Process

Depending on the agency and the size and extent of the project, the submittal and review process can either be brief or extend over a couple of years before a project is ready to be advertised for bids. This time frame should be clearly described in the A/E contract with the government agency. The A/E should review submittal requirements with the agency prior to starting a new phase of the work. During the design phase, the agency will often want reviews at milestones of 30 percent, 50 percent, 90 percent, and final completion. An A/E working with an agency for the first time may have a submittal rejected or found inadequate if procedures are not followed, and a resubmittal may be required. The following list outlines what may be expected from the A/E at each of the design reviews.

- 30 percent review, one of the following:
  - A list of specification sections intended for use on the project. The list should be generated from a master list of section titles furnished by the client agency.
  - An outline specification identifying products to be used.
- 50 percent review, one of the following:
  - A pencil or electronically marked-up copy of the complete agency guide specification, including specification pages that will be completely deleted.
  - A draft consisting of a pencil or electronically marked-up copy of the guide specifications, with unnecessary paragraphs deleted.
  - A typed or printed rough draft.

- 90 percent review, one of the following:
  - Resubmittal of the pencil or electronically marked-up copy of the complete guide specification, including all specification pages, even though some will be completely deleted.
  - Resubmittal of the draft consisting of a pencil or electronically marked-up copy.
  - Resubmittal of the typed rough draft.
  - Processed copy in final draft form.
  - One section in final form as a sample of the intended product.
- Final review:
  - Printed copy in final form.
  - If the reviewers have no comments, the project will be advertised for bids by the agency.

14.10.1 Review Comments

Agency reviewers often require changes to the drawings or specifications to make the documents conform to specific government criteria. The A/E should expect comments from the client-agency reviewers after each of the submittals. The A/E is expected to revise the contract documents and incorporate the review comments before the next submittal. There are several aspects of the review process:

- Each of the disciplines involves a different reviewer (e.g., architectural; structural; fire protection; plumbing; heating, ventilating, and air conditioning [HVAC]; and electrical), whose comments may result in conflicting requirements. The agency's project manager should resolve conflicting requirements among the various reviewers.
- Each reviewer's comments must receive a written response, usually on a form provided by the agency.
- Time schedules for agency review, A/E response time, and subsequent resubmittals should be documented in writing.
- Normally, a compliance-check submittal is required after major changes.
- Adequate time should be allowed in the contract document schedule to make revisions directed by the reviewers.

14.10.2 Printing for the Review Process

Depending on the terms of the agreement between the A/E and the client agency, either the A/E or the agency may be responsible for printing and distributing review copies of the contract documents. This can be an important item of both time and expense and should be considered when negotiating an agreement.

14.11 Regional Variations

Some federal agencies administer their construction programs through autonomous regional or district offices. For example, the COE publishes its guide specifications from its Washington, D.C., office, but the district offices often supplement and modify the specifications. In addition, a local military facility may contract directly with a local A/E firm for projects not subject to COE district jurisdiction. In such cases, the local military facility may have its own version of the guide specifications or may allow use of the A/E office master guide specification. The source of funding for a project may also dictate the procedures to be followed in executing the design contract. The A/E should be aware of possible variations in procedures.

Although each federal agency has its own procedures for administering design contracts, there are some similarities among various agencies:

- The use of section numbers based on *MasterFormat*® is common to most agencies.
- The A/E is prohibited from naming manufacturers or brand names for products, except in special circumstances.
- A number of agencies issue guide specifications for use by A/Es when preparing construction documents.
- A number of agencies have combined their resources to publish guide specifications in a common format, using SpecsIntact, and have provided distribution through NIBS CCB.
- Most agencies issue their own procurement requirements and contracting requirements and require that they be used in the project manual.
- Contract documents are subject to a review process, which can be lengthy and may require extensive changes and several resubmittals by the A/E.

Preparing specifications for federal government construction for the first time can be an involved and confusing experience if the A/E does not know what to expect. These difficulties are usually the result of having to follow rules and procedures that are somewhat different from similar work in the private sector. The general guidelines in this chapter may help the A/E anticipate and prepare for some of the differences likely to be encountered.

Chapter 15
Warranties

15.1 Warranty Overview

There are two basic types of warranties used in construction projects:

- Construction warranties or guaranties, which cover products and workmanship. Construction warranties are usually extended warranties specified in various specification sections.
- Warranties that cover products only.

Warranties on goods are normally affected by a sales transaction and are governed by the *Uniform Commercial Code* (UCC) and other applicable state laws, as well as by the specific terms of warranties offered. The UCC is a general term used to describe the law governing sales of goods and other commercial matters. Each state has adopted the UCC as the law in that state. Although the separate state enactments are similar, thereby providing uniform national law, there are some variations from state to state.

15.1.1 Definitions of Terms

The following definitions are taken from *Black's Law Dictionary* and apply to construction and purchasing warranties and guaranties:

Guarantor The person who makes a guaranty; one who becomes secondarily liable for another's debt or performance, in contrast with a strict surety who is primarily liable with the principal debtor.

Guaranty To undertake collaterally to answer for the payment of another's debt or the performance of another's duty, liability, or obligation; to assume the responsibility of a guarantor; to warrant.

Warrantor One who makes a warranty; any supplier or other person who gives or offers to give a written warranty or who is or may be obligated under an implied warranty.

Warranty

- A promise that a proposition of fact is true.
- A promise that certain facts are truly as they are represented to be and that they will remain so, subject to any specified limitations. In certain circumstances, a warranty will be presumed; this is known as an *implied* warranty.

Express Warranty Express warranties are created as follows:

- Any affirmation of fact or promise made that relates to the goods or installation and becomes a part of the basis of the bargain creates an express warranty that the goods or installation shall conform to the affirmation or promise.

- Any description of the goods or installation that is made a part of the basis of the bargain creates an express warranty that the goods or installation shall conform to the description.
- Any sample or model that is made a part of the basis of the bargain creates an express warranty that the whole of the goods or installation shall conform to the sample or model.

It is not necessary to the creation of an express warranty to use formal words such as *warrant* or *guarantee* or that the seller have a specific intention to make a warranty. An affirmation of the value of the goods or installation or a statement purporting to be merely an opinion or commendation of the goods or installation does not create a warranty.

Full Warranty A warranty as to full performance covering generally both labor and materials. Under a full warranty, the warrantor must remedy the consumer product within a reasonable time and without charge after notice of a defect or malfunction.

Implied Warranty A promise arising by operation of law that something that is sold shall be merchantable and fit for the purpose for which the seller has reason to know that it is required.

Unless excluded or modified, a warranty that the goods or installation shall be merchantable is implied in a contract for their sale if the seller is a merchant with respect to goods or installation of that kind. The serving for value of food or drink to be consumed either on the premises or elsewhere is a sale for this purpose.

Where the seller, at the time of contracting, has reason to know any particular purpose for which the goods or installation are required, and that the buyer is relying on the seller's skill or judgment to select or furnish suitable goods or installation, there is, unless excluded or modified, an implied warranty that the goods or installation shall be fit for such purpose.

Limited Warranty A written warranty that fails to meet one or more of the minimum standards for a full warranty. An example is a warranty limited to labor or to materials for a specified time. The UCC does not have a time limitation; thus manufacturers limit the time period.

Warranty of Title An implied promise that the seller owns the item offered for sale; the title conveyed shall be good, its transfer rightful, and the goods shall be delivered free from any security interest or other lien or encumbrance of which the buyer at the time of contracting has no knowledge.

15.2 Guaranties and Warranties

The terms *guaranty* and *warranty* are commonly used interchangeably to describe the responsibility of a manufacturer after delivery of a product and to describe the responsibility of a contractor after completion of construction. These same terms are also used for similar meanings in other contexts. According to *Black's Law Dictionary,* guaranty *and* warranty *are derived from the same root, and are in fact etymologically the same word, the "g" of Norman French being interchangeable with the English "w."*

Though the two terms are frequently misapplied in business, the law has assigned slightly different meanings to them. Legally, a guaranty is a separate contract by a third party (analogous to a surety bond), who assumes responsibility in case the principal fails to perform. Conversely, a warranty is assurance by the principal that it will assume stipulated responsibilities for completed portions of the project. Thus, a manufacturer warrants its material, whereas the construction contractor provides a third-party guaranty for those same materials and a warranty for the construction contractor's own workmanship in installing them.

15.3 Construction Warranties

The subject of warranties is as technical and confusing as that of insurance. Although most architect/engineers (A/Es) have learned to avoid specifying insurance requirements, it is still common practice for A/Es to specify warranties. Those who do so on behalf of the owner must be as knowledgeable about warranties as they are about the physical properties of the products and materials they specify.

15.4 Terms and Conditions

Requirements that apply to maintaining valid warranties are referred to as *terms and conditions.*

The *American Institute of Architects (AIA) A201 General Conditions of the Contract for Construction*, the *Engineers Joint Contract Documents Committee (EJCDC) C-700 Standard General Conditions for the Construction Contract,* and other Standard General Conditions as well as derivative documents, indicate that the contractor warrants that material and equipment furnished is new, or good quality, free from defects, and conforms to the requirements of the contract documents. These warranty provisions go on to indicate exclusions of defects and damage caused by insufficient or improper operation and maintenance, abuse, modifications, and normal wear and tear. It should be noted that none of these warranties include a time limitation. The time limitation will vary depending on the location of the project and the laws governing the contract. Although exclusion limits the applicability of a warranty, a condition must be satisfied in order to make the warranty effective. If the condition is not met, the warranty is null and void. In the case of a roofing warranty, some requirements, such as proper use and adequate maintenance, may be stated as conditions rather than as exclusions. In such cases, the manufacturer may disclaim liability even if the product failure has nothing to do with the failure to comply with the condition. There is a widely held misconception that a contractor's responsibility for defective work lasts only until the end of the one-year correction period. However, the courts have consistently held that expiration of the guaranty period does not terminate the contractor's responsibility for defective work[1] except for statute of limitations.

[1] *John W. Cowper Co. v. Buffalo Hotel Dev.,* 115 A.D. 2d 346, 496 N.Y.S. 2d 127 1985; *Omaha Home for Boys v. Stitt Construction Co., Inc.,* 238 NW 2d 470, 1976; *Baker-Crow Construction Co. v. Hames Electric,* 566 P2d 153, 1977; *City of Kennewick v. Hanford Piping,* 588 P2d, 1977

In an attempt to clarify this misconception, the AIA, EJCDC, Design-Build Institute of America (DBIA), Associated General Contractors (AGC), and other producers of standard contract documents have terminology in their standard general conditions that refers to *correction period* and do not use the term *warranty period.*

This correction period is not a limitation of the warranty provisions of the general conditions.

During the correction period, the parties assume that responsibility to correct failures rests with the contractor. After the end of the correction period, the parties assume that responsibility to correct failures rests with the owner. In the law, an assumption of this type is called a *presumption.* This presumption can be refuted but will be followed unless evidence is presented to overcome the presumption.

The one-year correction period is simply a specific contractual obligation for the contractor and a remedy for the owner under the general contract. It is not intended to limit the effect of warranties provided in or required by the contract. The distinction between the obligations of the contractor under the general warranty and its one-year specific correction obligations is not always made. Some contracts, most notably federal construction contracts that use warranty language provided in Federal Acquisition Regulations (FAR), actually limit the contractor's warranty obligations to a one-year period.

Extended warranties are sometimes required in the specifications covering items such as roofing. Unfortunately, many of the documents that purport to be guaranties or warranties are actually little more than disclaimers of responsibility. It is virtually impossible to obtain a manufacturer's warranty that is as broad as the responsibilities that a construction contractor assumes during the correction period. Figure 15.1 illustrates a two-party agreement with contractual assurance by the contractor to the owner of a warranty obligation for a specified time period. With use of AIA, EJCDC, and other standard general conditions it is unnecessary to specify one-year warranties. Extended warranties are not provided for in AIA, EJCDC, and other standard general conditions and must be appropriately specified in PART 1—GENERAL of the product specification section.

Figure 15.2 illustrates the relationship of the owner and the contractor and the requirements of a guaranty document similar to an insurance policy or a performance bond and involvement of a third party (guarantor or surety) to that contract.

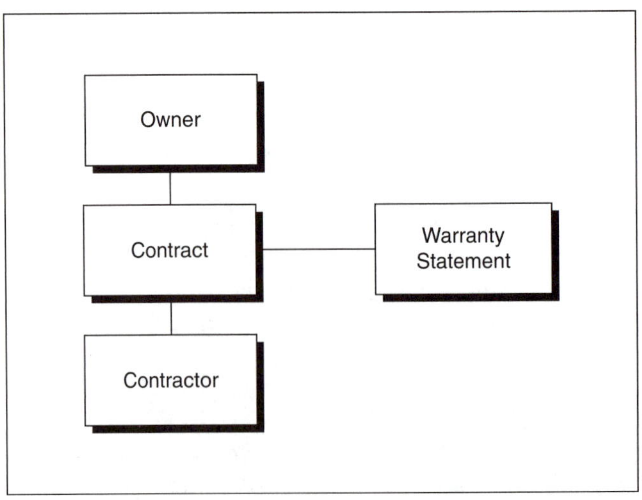

Figure 15.1
Example of a Two-Party Warranty Relationship

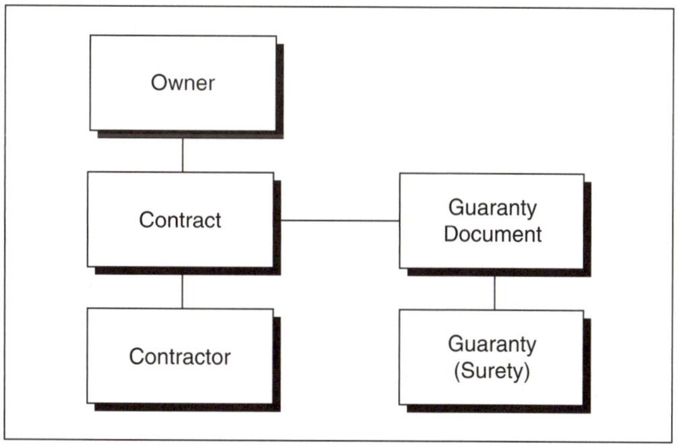

Figure 15.2
Example of a Three-Party Guarantor Relationship

15.5 Purpose

Construction warranties are usually required for several reasons:

- To protect the owner against faults, defects, or failures, in spite of technical compliance with the terms of the contract
- To provide a remedy to the owner for nonconformance with the contract after completion and acceptance of construction
- To give the owner recourse against additional parties (manufacturers, subcontractors, and suppliers) who are not in a direct contractual relationship with the owner
- To extend the manufacturer's responsibility beyond the end of the correction period
- To allow a remedy beyond the normal statute of limitations

The unpredictability of performance and replacement cost of some products, such as roofing, waterproofing, insulating glass units, compressors, and other equipment, causes many owners to insist on extended warranties. The A/E should carefully evaluate the decision to specify warranties on the basis of the cost/benefit to the owner.

Many warranties are offered because of market conditions or as marketing techniques. Until just a few years ago, the built-up roofing industry offered a third-party *roof bond,* which was considered of little value. However, when the single-ply roofing systems became available with extended warranties, the built-up roofing industry began offering similar warranties in order to remain competitive. In some cases, when performance and other requirements of two products are similar, the decision to purchase a particular product may be based on the available warranty.

In order to introduce a new product, some manufacturers will offer an extended warranty as an incentive to have their product specified or accepted. Extended warranties also may be offered on products that do not quite measure up to the project specifications.

15.6 Benefits of Extended Warranties

For anyone to benefit from extended warranties, the manufacturer must be financially secure enough to cover its liabilities. Immediate benefits can be gained from an equitable extended warranty that could make it worth the cost to the owner. These benefits include the qualification of the installer by the manufacturer, the manufacturer's involvement in the construction process, and insurance against failure.

One way to ensure these benefits is to specify and purchase the manufacturer's best system, which, for built-up roofing, often means a 20-year system. When A/Es specify systems with shorter warranty durations, they may find that, in some cases, roofing contractors can become approved installers simply by placing a phone call to the manufacturer. However, when a 20-year system with a full workmanship and materials warranty is specified, most manufacturers prequalify the installer and get involved in the construction process. The 20-year warranty means that the manufacturer has a vested interest in supplying a long-life roofing system that should provide trouble-free service when properly maintained by the facility owner.

15.7 Limitations and Exclusions

Product warranties are frequently perceived as providing increased legal and financial protection to owners against product defects. This is usually an inaccurate perception. Most product warranties provide a limited warranty, which actually reduces the rights an owner may have by statute under a full warranty. Now that strict product liability has become a major concern for manufacturers, the terms of warranties have become a principal mechanism for limiting manufacturers' liability for products that are defective or fail to perform as expected.

When an owner decides to insure a facility, the insurer can evaluate its risk in terms of geography, facility type, facility contents, quality of construction, and so on. This can be done because a structure exists to evaluate. However, when a manufacturer sells a product, little is known of the exact circumstances under which its product will be required to perform. Therefore, a manufacturer can either limit its potential liability or choose not to and accept the risks associated with its decision. Generally, manufacturers choose to limit their exposure and price their products accordingly.

Unless otherwise stipulated, materials are sold under a full warranty. According to the UCC, a full warranty includes not only refund or replacement for the defective product, but also the labor to repair or replace the product, as well as "consequential and incidental damages" that occur as a result of the failure of the warranted product. These damages could include damages to furnishings and adjacent construction, and even profits lost by the owner because of damaged merchandise or the inability to conduct business under normal conditions. The cost of replacement and consequential and incidental damages can easily exceed the original cost of the defective product. For example, a waterproofing material that fails to perform as warranted may cost less than a dollar per square foot, but it could cost thousands of dollars to replace and to repair damage in areas located below the waterproofing.

Most manufacturers find that the risks of a full warranty are more than can be accepted without increasing the cost of their product to uncompetitive levels. But the UCC allows a buyer and seller to agree to a limited warranty with reduced remedies. There is seldom a real negotiation on the terms of a product manufacturer's warranty. Instead, a manufacturer typically disclaims its obligations to provide remedies under a full warranty and incorporates a limited warranty into its terms of sale. If the disclaimers are conspicuous in the product literature and on the product labeling, then specifying or purchasing a product will normally imply the buyer's consent to the warranty limitations. Limited warranties also seek to avoid litigation by establishing in advance the terms and remedies available under the warranty.

Common limitations in a warranty can exclude labor and consequential and incidental damages. Other conditions may establish criteria that must be met by the buyer before the warranty will be honored. For example, a manufacturer may require an extra fee to register the warranty, restrict the conditions of use for the product, or require installation by specially qualified contractors. Other provisions may limit rights under the warranty to the original purchaser. This can be interpreted to mean a contractor or a subcontractor, leaving an owner without recourse directly to the manufacturer. If the product warranty begins when the contractor purchases or installs the product, the owner will receive less protection than if the warranty began at substantial completion. The A/E should study the disclaimers and limitations to determine what protection is provided under a warranty and which owner's rights and remedies are taken away.

The A/E should also verify that warranty provisions comply with code requirements. As an example, one major roofing manufacturer's standard warranty excludes responsibility for damage caused by "winds of peak gust speeds of 55 mph or higher measured at 35 feet above the ground." However, model building codes might list 70 mph as the minimum design sustained wind load and 85-mph three-second gusts.

Though a full warranty (materials and labor) is specified, there may be limitations that affect the warranty. In the case of waterproofing, for instance, the cost of removal and replacement of other materials that cover the waterproofing is generally not included in the manufacturer's warranty, or a sealant manufacturer may offer a 20-year warranty that is valid only if the sales receipt and original container are returned. Some warranties may cover replacing the initially installed product but not cover the replacement product itself. This will leave some parts of a project without the full extended warranty. Some products cannot be field modified without voiding their warranty, and others have strict limitations on how they may be modified. The financial capability of the manufacturer or supplier offering an extended warranty should always be considered. Warranty forms should be requested along with other submittals to verify all limitations, terms, and conditions.

As extended warranties are usually written from the manufacturer's perspective, even those issued in good faith invariably contain language that limits the scope of their coverage. Warranties usually exclude consequential damage to any facility component other than the warranted product itself. In many cases, coverage of a product or system is prorated; for each year of service the product or system has already provided, many warranties pay a smaller percentage of the repair or replacement costs.

When manufacturers delete exclusions from their warranties, they usually add other costs or terms and conditions to protect their interests. Prorated coverage, for example, is eliminated in *no-dollar-limit* roof warranties. In exchange for paying full replacement costs for a defective system, the manufacturer requires the use of all or almost all of the manufacturer's products within a warranted system.

Given the number of ways that a manufacturer's obligations can be diminished or nullified, it would be a mistake to place marginal systems on a level with their more expensive competitors simply because both systems are warranted for the same number of years. In reality, the terms of one warranty may provide significantly more protection than the terms of another.

A/Es should carefully read warranties and determine what is covered and what is excluded. The exclusions are very important. Some roofing warranties exclude damage caused by ponded water and gale winds. Many manufacturers' warranties take away consumer protection normally included in the UCC, including the implied warranty of merchantability and fitness. A manufacturer's warranty may restrict repairs of failures only up to the cost of the original installation, not replacement costs. An extended warranty is of little benefit if nothing is covered. For example, an implied warranty becomes worthless if the written warranty states that a roofing material may be unsuitable for roofing.

The only way to be sure that warranty dollars are being invested wisely is to compare the clauses and conditions of several warranties and identify those that could cause problems if the need to file a claim arises. For instance, clauses stating that the terms of the warranty are the owner's sole remedy nullify any added protection the owner would have had through implied warranties.

In addition to exclusions and limitations, the A/E should examine a manufacturer's ability and willingness to honor the manufacturer's warranties. An important factor is the financial backing of the warranty under consideration. Some manufacturers do not have or have not committed sufficient assets to satisfy all claims against their warranties. In either case, an owner may be left with a faulty product and no immediate remedy.

It is also important to determine which companies have been in business long enough to have long-term experience with their products. A 20-year warranty issued by a company that has been in business for 5 years is questionable, just as a 20-year warranty on a new product is also questionable. In some cases, it may be useful to talk to facility owners with warranted systems from a specific manufacturer to determine its product history. Reports of continual repairs and slow or negative responses to warranty calls may indicate a problem with a product or system, or it may indicate that a company is in financial trouble. Sometimes the warranty provider is a broker, not the manufacturer. The A/E should determine whether the warrantor is the same company that has provided the product or a broker. Also, the A/E should determine whether the warranty is financially backed by the product manufacturer.

The A/E should be aware of the limitations warranties place on repairs or alterations of warranted items. Usually, a repair or alteration to an item by a party other than the manufacturer or original installer will invalidate the remainder of the warranty period.

The following exclusions and limitations should be of particular concern:

- *Language Making the Warranty the Exclusive Remedy.* These clauses reduce the possible remedies that the owner could seek if there is a problem.
- *Clauses Limiting the Scope of Coverage.* A/Es should ask if the warranty covers materials only or workmanship as well.
- *Clauses Limiting the Assignability of the Warranty.* Generally, the law will assume that a warranty is transferable if the facility is sold before the warranty expires. If the warranty specifically prohibits assignment and the facility is sold, the new owner is not covered by the warranty.
- *A Requirement Stating That the Owner Must Sign the Warranty Document.* If the warranty is not effective until it is signed, there is a good chance it contains language that will limit the owner's rights.
- *Warranties Containing a Deductible.* If included, determine when the deductible begins.
- *Clauses Limiting the Time That the Owner Has to Take Legal Action* for breach of a warranty.
- *A Requirement Allowing the Warrantor to Recover Legal Costs* in a proceeding involving the owner or contractor.
- *Unfair Dispute Resolution Procedures* when a difference in opinion occurs regarding the manufacturer's warranty.
- *Clauses That Exclude Installation of the Replacement Product.*

Warranties are only as good as the manufacturer's reputation and become useless if the manufacturer goes out of business. Bonds may be considered for major items if extended warranty periods are required. Also, an insurance policy with coverage similar to the required warranty may be an acceptable alternative. The owner's insurance adviser should be consulted if this alternative is considered. Some warranties provide that the warrantor reserve the exclusive right to determine whether the warranty is applicable; in such a case, the A/E should try to determine whether a mechanism is available to appeal the rejection of a claim.

It is conceivable that a comparison of manufacturers and their warranties could lead to selection of a product or system that has a good performance record but a very limited warranty. Unfortunately, many manufacturers are reluctant to negotiate the terms of their warranties, so an owner's options may be limited to accepting unfavorable conditions, using another product or system with an acceptable warranty, or choosing not to accept any manufacturer's warranty and relying solely on the provisions of the UCC.

Ultimately, an owner may accept an extended warranty on the manufacturer's terms. However, in the process of examining and comparing warranties, the owner will have learned what level of service can realistically be expected. More important, the owner will have avoided the trap of assuming that a product's expected service life is equal to the length of its warranty.

The A/E should address the problem of unwanted exclusions by listing acceptable exclusions in the specifications but requiring the A/E's approval of additional exclusions as a condition of acceptance of the product or its warranty. Specifications that generally require the contractor to "install in accordance with the manufacturer's instructions" will help to preserve the owner's rights under the warranty, but only if the contractor complies. Despite their limitations, warranties do instill buyer confidence in a product.

15.8 Remedies

The UCC permits the manufacturer to disclaim implied warranties of merchantability and fitness in writing, and also permits vendors to limit the duration of their responsibility as well as the remedies that are available to the buyer. The nature of those remedies and their limitations are the most important features of warranties.

The A/E who requires only that the contractor submit the manufacturer's standard warranty could be giving the manufacturer the opportunity to limit its responsibility to a level that is unreasonable from the owner's standpoint. The owner may be better off having no warranty at all.

The time it takes to settle a claim can also cause problems. Most manufacturers have a procedure that contractors must follow to collect for repairs made under warranty. If the contractor proceeds immediately to make repairs without waiting for the manufacturer's approval, the repairs may not be compensated by the manufacturer. This can place contractors in a difficult situation. Contractors may jeopardize long-standing business relationships if they do not respond quickly to an owner's request to repair a defective product. When the owner is a large company with many facilities that require some sort of repair work, contractors should not jeopardize future work by being unresponsive in emergency situations.

In the case of a multi-component system, the A/E should require the manufacturer issuing the warranty to accept system responsibility. The A/E should specify a warranted system, not a system with individually warranted components. When problems occur, the owner should not have to rely on the legal system to determine who is responsible for a product or workmanship failure. Because of the association with remedial provisions, warranties are often thought of as a prescribed set of remedies promised by the warrantor, rather than a representation of the condition or performance of the warranted item. Some warranties may be prorated over the stipulated period of time and thus may diminish in value over the duration.

15.9 Duration

The most commonly specified factor for warranties and the most common basis for comparison of warranties is their duration. However, the duration actually may be one of the least important terms of the warranty, and the failure to address some of the other terms may leave the owner without the desired protection. By specifying only the duration for a warranty with no further requirements, the A/E will have no basis for rejecting a warranty that has the specified duration but is unacceptable in other respects.

Express time limits are generally stipulated in product warranties. The time limits often bear no direct correlation with the expected service life of the product, and in some cases it is unrealistic to expect a product to serve satisfactorily for the duration of its warranty. Rather than relying on a warranty, the A/E should verify the performance history of a particular product before specifying it for an application where duration of performance is critical. Another area worth analysis involves the extent of testing or other studies conducted by the manufacturer to verify that the system can reasonably be expected to function for the stated period.

15.10 Purchasing Warranties

Warranties effected by a sales transaction and governed by the UCC and other applicable state laws are referred to as *purchasing warranties.* Although purchasing warranties are seldom directly applicable to a construction contract, the UCC governs most transactions involving the sale of goods. The UCC attaches certain express and implied warranties to sales transactions and governs warranties provided by the seller. Any written warranty is called an express warranty. An implied warranty is derived from the nature of a transaction, applicable state law, and the position of the parties.

Express warranties may consist of:

- An affirmation of fact or promise related to goods, such as claim of performance
- A description of the goods, such as photos, drawings, or specifications
- A sample or model that becomes a basis of the agreement

Under the UCC, there are two implied warranties: the first is the implied warranty of merchantability or "implied warranty that goods are merchantable," which means that the consumer goods:

- Pass without objection in the trade under the contract description
- Are fit for the ordinary purposes for which such goods are used

- Are adequately contained, packaged, and labeled
- Conform to the promises or affirmations of fact made on the container or label

The second is the implied warranty of fitness. A warranty of fitness means that when the retailer, distributor, or manufacturer has reason to know any particular purpose for which the goods are required and, further, that the buyer is relying on the skill and judgment of the seller to select and furnish suitable goods, then there is an implied warranty that the goods shall be fit for the intended purpose.

These implied warranties are a part of every transaction unless the sale documents clearly state that they do not apply. The implied warranty of merchantability requires that the goods be reasonably fit for the general purpose for which they are sold. The term *merchantable* means fair, average, medium quality.[2]

Regarding the implied warranty of fitness, *Smith and Roberson's Business Law* states:

> Any seller ... impliedly warrants that the goods (he sells) are reasonably fit for the particular purpose of the buyer for which the goods are required, if at the time of contracting the seller has reason to know such particular purpose and that the buyer is relying on the seller's skill and judgment.

To illustrate the difference between the two types of implied warranties, compare a specification calling for a pump with one calling for a submersible pump. The seller of the pump impliedly warrants that it is fit (merchantable) for the ordinary purposes for which pumps are used. The seller of the submersible pump might also warrant that it will function under water (fitness for a particular purpose).

The UCC also governs the manner in which warranties may be disclaimed. Exclusion of the warranty of title must be by specific language, or the circumstances of the sale must be such that the buyer should not expect a warranty of title. Express warranties can be disclaimed in the same manner in which they are created, but warranty language will take precedence over disclaimer language in the event of a conflict. Implied warranties may generally be disclaimed by the use of a *conspicuous* written disclaimer or by the use of the words "as is, with all faults" or similar language. Federal law requires that written warranties covering consumer products (often inapplicable to construction projects) fully and conspicuously disclose, in simple and easily understood language, the terms and conditions of such warranty, including whether the warranty is a full or limited warranty.

To obtain maximum advantage of the implied warranty of fitness, sellers should be informed of the specific uses to be made of their goods and that their recommendations are being relied upon. If the buyer requests or demands a particular brand, there is no implied warranty of fitness because the buyer is relying on the buyer's own skill or judgment.[3]

When specifying warranties, the A/E should avoid:

- Relying on a warranty as a substitute for thorough investigation of a product and its manufacturer
- Requiring warranty coverage that is not available for a particular product
- Requiring or permitting a warranty that weakens, rather than strengthens, the owner's rights

[2] Richard A. Mann, Barry S. Roberts, and Len Young, *Smith and Roberson's Business Law,* 12th ed., Southwestern, 2002.

[3] Mann, Roberts, and Young. *Smith and Roberson's Business Law.*

The A/E should be as familiar with standard warranties offered with products as with the physical qualities and performance characteristics of the products. The size, stability, and reputation of the manufacturer and its ability to make good on a warranty claim may be more important than the terms of the warranty itself.

The A/E should determine before producing the specification whether warranties are offered or are available, the terms of such warranties, whether the terms can be specified, whether the terms of standard warranties are negotiable, whether the owner can do without the warranty altogether, or whether the manufacturer requires acceptance of its warranty terms as a condition of sale.

In general, the best assurance of long-term product performance is a combination of good materials, appropriate design, and skillful workmanship. No warranty can compensate for the lack of these important aspects.

Chapter 16
Construction Bonds

16.1 Construction Bond Overview

There are three basic types of bonds used in construction projects: bid bonds, performance bonds, and payment bonds. The first, if used, is a procurement requirement; the latter two are classified as contract documents. Other types of bonds may be required, such as warranty bonds or maintenance bonds, but these are generally called for by individual specification sections.

As the author and coordinator of the project manual, the architect/engineer (A/E) must be familiar with the use of bonds on construction projects. Even though it is the owner's decision on whether to require bonds, prior to the decision being made the A/E should confer basic information regarding the use and form of bonds with the owner. Regardless, it is imperative that the A/E recommend the owner seek bond advise from their counselor or attorney, and a professional surety advisor.

16.2 Definitions

A surety bond is a three-party contract under which one party promises to answer for the debt or default of another. The critical terms used in bonding include the following:

Principal The party who has the primary obligation to perform the undertaking that is being bonded. For example, the contractor on a bonded construction project is the principal.

Surety Also referred to as the bonding company, the surety is the party that guarantees the principal's performance. In essence, the surety agrees to be bound to the obligations of the principal should the principal fail to perform them.

Obligee The obligee is the person for whose benefit the bond is written. With respect to a performance bond, the obligee is usually the owner. Some performance bonds are written in favor of more than one obligee (e.g., a subcontractor's bond may be written for the benefit of both owner and contractor, or a contractor's bond for the benefit of the owner and the project lender). Such bonds are called *dual obligee* bonds.

Surety Bond The surety bond is the written document given by the surety and principal to the obligee to guarantee a specific obligation.

Indemnity Agreement An agreement between the principal and the surety whereby the principal guarantees the surety that the surety will incur no loss by reason of its providing the bond.

Penal Amount Bonds are written with a limit on the amount of the guarantee. This limiting amount, frequently 100 percent of the contract amount, is called the *penal sum* or *penal amount* of the bond.

Claimant Commonly used to refer to a party who files a claim against the bond. This could be the owner or a subcontractor or supplier seeking recovery under the payment bond.

Lien The legal right of a party, such as a subcontractor, to claim a security interest in the project or have it sold for payment of a claim.

16.3 Purpose of Bonds

Bonds generally provide the owner, subcontractors, and suppliers with assurance that should the contractor fail to make payments or complete the contract, there will be a financially solvent party, the surety, to perform these obligations. A surety bond is not an insurance policy; rather, bonding involves a three-way contractual relationship among the principal, surety, and obligee. A surety's obligation is analogous to the obligation of a cosigner on a note. The principal remains primarily liable to perform the bonded contract and is obligated to indemnify the surety against any loss.

The surety will normally have a separate indemnity agreement with the principal, and if the principal is a closely held corporation, with the individual owner, officers, or operators of the principal. Under the indemnity agreement, the principal and any indemnitors guarantee that the surety will incur no loss by reason of providing the bond. The principal pays a premium to the surety for providing the performance and payment bonds.

The premium is based on the amount of the contract being bonded and normally is between 0.3 and 2 percent, depending on various factors including the type of risk undertaken, financial strength of the principal, financial strength of any indemnitors, past claims history, and extent of the project. Bid bonds are usually offered at no cost if performance and payment bonds are required. It is important to understand the three-party relationship involved in these bonds (see Figure 16.1). Unlike

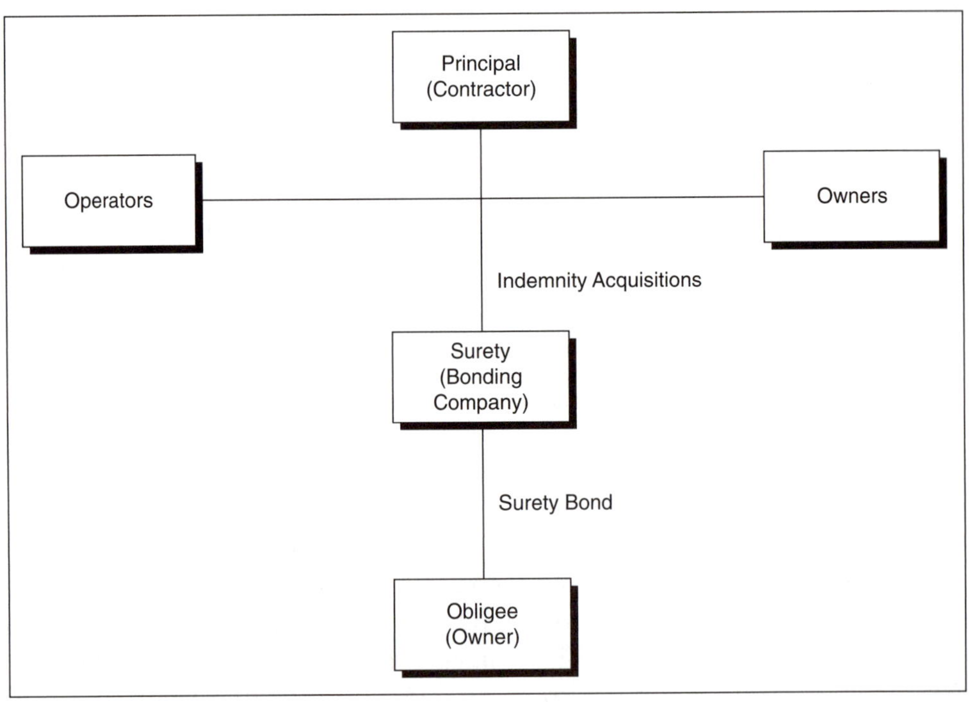

Figure 16.1
Construction Bonding

insurance policies, where the insurance company accepts primary responsibility to pay claims, the surety is only secondarily liable after the principal defaults. The surety theoretically should not expect to make any payment out of its own pocket without reimbursement from the principal or individual indemnitors. The surety should be licensed by the insurance department of the state where the project is located. Offers by a contractor to provide bonding through an individual or an unlicensed surety should generally be rejected.

16.4 Rating of the Bonding Company

The A.M. Best Company rates insurance companies. *Best's Key Rating Guide* is published annually, with rating classifications that range from an A+ for a superior rating to a C- for a fair rating. An A- or better rating suggests that the carrier is financially excellent. Standard & Poor's, Fitch, and other companies also publish ratings of insurance companies.

The federal government has its own criteria for sureties. The U.S. Treasury Department publishes an annual listing of companies holding certificates of authority as acceptable sureties on bonds provided to the federal government. Carriers on the federal Treasury's Listing are acceptable for federal government projects. The Treasury's Listing is published annually in the July 1 *Federal Register* and is available on the Treasury Department web site at http://fms.treas.gov/c570/c570.html.

16.5 Obtaining Bond Information from the Owner

The A/E usually completes the project manual prior to issuing a project for bidding or negotiation, which includes the procurement requirements and conditions of the contract. The proper place to place bid bond information is in the procurement requirements. The supplementary conditions should state the requirements for performance and payment bonds. This information should be requested from the owner early enough to allow the owner to respond to the A/E in sufficient time for the information to be incorporated into the project manual. When performing work for a repeat client or with an owner constructing multiple projects, the A/E should request bond information for each individual project.

Bond information request forms published by either the American Institute of Architects (AIA) or Engineers Joint Contract Documents Committee (EJCDC) are recommended for documenting the bonding requirements. These forms provide the A/E with a standardized means of ensuring that the information has been requested and obtained in the proper manner. The AIA form is Document *G612, Owner's Instructions to the Architect Regarding the Construction Contract.* The EJCDC form is *Document C-052, Owner's Instructions to Engineer Concerning Bonds and Insurance for Construction.* In addition, EJCDC also publishes a prototype letter for requesting the information. This document, *EJCDC C-051*, is titled *Engineer's Request for Instructions Concerning Bonds and Insurance for Construction.*

16.6 Advantages and Costs of Bonding

There are numerous advantages of providing a fully bonded project:

- The bid bond protects the owner against the withdrawal of a favorable bid.
- The suppliers and subcontractors are protected against nonpayment of amounts due them. The owner is protected against mechanic's liens on the project.
- The owner is protected against default, breach of contract, and nonperformance by the contractor.
- The owner receives additional assurance of the stability of the contractor. Most bonded contractors are financially stable, and the bond guarantees that the surety will pay if the contractor fails.
- The indemnity agreement generally provides added incentive for the principals of the contractor to properly perform, as they may be personally liable to the bonding company for amounts paid on a bonded project.
- Bonding satisfies statutory requirements for publicly funded projects.

There is a cost for this protection. The bond premium will be included in the contractor's bid, and one of the functions of a bond requirement is to exclude contractors who cannot qualify for bonding. For residential buildings and small commercial contracts, some potential contractors may not have established relationships with sureties and be discouraged from bidding by a bond requirement.

The decision whether to require bid, performance, and payment bonds is one that should be made by the owner. However, owners often ask A/Es to provide information, input, or forms concerning bonding for a construction project. Unless the owner is specifically cautioned otherwise, the statements of the A/E may be accepted without further investigation, especially when owners do not have separate legal counsel or insurance advisers. If bonds are not required and a contractor defaults, the owner may well contend that the A/E's advice concerning bonding was insufficient. Although an A/E may provide some basic information to an owner, owners should always be advised to seek the counsel of an attorney or surety adviser before making a final decision.

16.7 Bid Bond

The purpose of the bid bond is to protect the owner from losing the benefit of an accepted bid. Issuance of a bid bond commits the bidder to enter into a contract and, if required, provide performance and payment bonds. The bid bond is provided by the bidder at the time of bid and is submitted with the bid. The bid bond generally provides for a penal amount expressed either in dollars or as a percentage of the total amount of the bid. In the event that the selected bidder fails or refuses to enter into a contract for the price that was bid, the surety is obligated to pay the owner's damages, up to the penal amount of the bid bond. When bid bonds are required, failure of a bidder to submit a bid bond will make such bid nonresponsive and result in rejection of that bid. Bid bonds are generally cost-effective for the owner, as most bonding companies do not charge separately for the bid bond if a performance bond and a payment bond are also required.

When a selected bidder finds that its bid is substantially below that of the other bidders or that a mistake has been made that is not subject to correction or modification, the accepted bidder may be reluctant to enter into the contract in that amount. If the owner requests that the selected bidder sign a contract in the amount of the bid and the selected bidder refuses, the owner can go to the surety and ask for the difference between the amount of the principal's bid and that of the next lowest bidder, up to the penal amount of the bid bond. Thus, the owner will still get the benefit of the low bid. Further, if the selected bidder does not sign the agreement for some reason and the owner is unable to enter into a contract with the next lowest bidder, as might be the case if the other bidders have already withdrawn, the owner may have to rebid. In that event, the surety not only is generally responsible for any difference between the principal's bid and the low bid received on the re-procurement, but may also be liable for the cost of rebidding.

The bid bond is liable only if the principal's refusal to sign the contract at the bid price is unjustified. If the principal was entitled to withdraw its bid, then there is no liability under the bid bond. For example, the bid bond could be discharged if the bid was not accepted within the time specified in the procurement documents or the bidder proves a mistake that entitled it to withdraw its bid or the owner tries unilaterally to change the terms and conditions under which the bid was submitted.

16.8 Performance Bond

The performance bond provides the most important protection for the owner by guaranteeing that if the contractor defaults, the surety will either complete the contract in accordance with its terms or provide sufficient funds, up to the penal amount of the bond, to fund such completion.

The AIA A312 performance bond form and the comparable EJCDC form require the owner to request that the surety attend a predefault meeting at which an attempt can be made to prevent default. Often, financial reinforcement by the surety will permit the contractor to complete the work. If there is a default and the owner makes a demand on a surety to perform under the terms of the performance bond, the surety will investigate the owner's claim against the contractor. This investigation will include examining both the owner's and the contractor's positions and making an independent analysis of the situation. Once the validity of the claim has been established, the surety has a number of options as to how it will fulfill the bond obligation.

If the contractor's work is satisfactory and the problems are only financial, the surety may choose to finance the contractor to complete the project. Under this option, the surety provides funds to the contractor so that the contractor can pay subcontractors and suppliers, buy materials, make payroll, and take other steps necessary to complete the project. The surety normally does this under careful supervision.

If the owner terminates the contract, the surety can complete the project by taking over the work itself and hiring another contractor, or by arranging for a completion contractor to work directly for the owner. The surety may also leave it up to the owner to finish the project and pay the difference between the balance left in the original contract and the actual cost of completion. If there is a dispute concerning the contractor's default or the surety's liability, the owner may have to proceed with completion of the project and then prove its claim in litigation or arbitration before receiving payment from the surety.

Whichever method of completion is used, the surety is entitled to credit for the contract balance at the time of default and is liable only for the excess costs to complete.

The performance bond protects the owner against default by the contractor, not against the owner's own defaults. If the owner materially breaches the contract and discharges the contractor, or if the owner breaches the contract and increases the surety's risk, the surety may have defenses to any claim on the bond.

For example, if the owner pays the contractor more than is owed under the contract or materially increases the scope of the contract without the surety's consent, the surety can claim a discharge to the difference from the original amount.

16.9 Payment Bond

The payment bond, sometimes referred to as the labor and materials payment bond guarantees that subcontractors, material suppliers, and others providing labor, material, and equipment to the project will be paid. This promise benefits the owner because it protects against mechanic's liens and delays caused by unpaid subcontractors and suppliers. The payment bond generally provides for payment not only to parties employed by or in direct contractual relationship with the contractor but also to sub-subcontractors and suppliers to subcontractors. Some bonds and statutes may include even more remote claimants within the coverage of the payment bond. To alleviate the unfairness of making the contractor pay remote claimants when the first-tier subcontractor has already been paid for the work, most bonds require such remote claimants to give the contractor written notice that they are unpaid within 90 days of the last date they furnished labor or material to the project.

Payment bonds also typically require that a claimant file suit within a reasonably short time. Only labor and material on the bonded project are covered, and the purchase price of capital goods or inventory available for use on multiple projects is not generally an obligation of the bond.

16.10 Bond Forms

In federal, state, county, and many municipal projects, the A/E will have no role in the form of the bond because the language is standard and, in some cases, required by law or statute. However, in situations where there is no standard format, the A/E should advise the owner to seek legal counsel regarding the choice of bond forms.

EJCDC has two published bid bond forms, *EJCDC C-430, Bid Bond, Penal Sum Form*, and *EJCDC C-435, Bid Bond, Damages Form*. EJCDC C-430 is the traditional bid bond form in which the owner is paid the face amount of the bond if the selected bidder fails to execute the owner-contractor agreement. EJCDC C-435 is slightly different in that the owner is paid the actual amount of damages upon failure of the selected bidder to execute the agreement, up to the penal sum of the bond. Usually, the damage is the difference between the bid of the defaulting bidder and the bid of the next-lowest responsible bidder. *AIA Document A310, Bid Bond*, is another widely used bid bond.

AIA and EJCDC each include performance and payment bond forms in their contract documents packages. Construction performance bonds are AIA Document A312 and EJCDC C-610. Construction payment bond forms are AIA Document A312 and EJCDC C-615. Other producers of standard and nonstandard contract documents also produce related bond forms. The bond forms selected for use in a project should coordinate and comply with local jurisdictions, and any related language would reference the project documents. There are important legal consequences in the choice of bond forms. If a form is not included in the procurement documents, the form of bond may become a matter for negotiation between the owner and the contractor/surety. This situation can be avoided by including the required bond forms in the procurement documents.

Evidence of authority to bind the surety is usually in the form of a power of attorney designating the agent authorized to sign on behalf of the surety. The instructions for procurement should require that a copy of the power of attorney be filed with the signed and executed bond form.

Chapter 17
Construction Insurance

17.1 Construction Insurance Overview

This chapter covers basic insurance terms, some of the risks involved in construction projects, and the kinds of construction insurance policies available. It does not address what constitutes adequate insurance coverage or insurance requirements for projects outside the United States. The owner's legal counsel and insurance adviser should be the source of all recommendations on construction insurance matters. The architect/engineer (A/E) should not act as an insurance adviser and should not determine or recommend insurance limits. However, the A/E should provide the owner with American Institute of Architects *(AIA) Document G612, Owner's Instructions to the Architect Regarding the Construction Contract*, or *Engineers Joint Contract Documents Committee (EJCDC) C-052, Owner's Instructions Concerning Bonds and Insurance for Construction*, both of which are insurance information request forms to obtain instructions for incorporating insurance requirements into the supplementary conditions.

17.2 Insurance for the Construction Contract

The owner of a construction project risks substantial assets and is exposed to significant legal liability throughout the life of a project. The owner also has a significant amount of money invested. The contractor performing the work also incurs risks that require insurance protection for itself, the owner, and third parties.

The owner's first line of financial defense is an adequately insured contractor. The owner's interest is protected in part by requiring the contractor to carry specific types of insurance coverage, within limits determined to be adequate for the circumstances. The insurance coverage required by the conditions of the contract provides most of the protection needed for both the owner and the contractor during the course of a construction project. However, such contractor's insurance may not provide all the coverage the owner needs. A number of policies are available to owners for coverage beyond the protection provided by the contractor's insurance.

17.3 Liability Insurance

Liability policies protect the named insured from losses arising out of legal liability to others caused by the insured's activities. Liability policies do not cover damage or loss of the insured's products, machinery, or equipment. Coverage is effective only at the project site.

Both AIA and EJCDC general conditions require that the contractor provide liability insurance because it is the contractor who is responsible for activities at the project site. Figure 17.1 shows the relationships among the different types of liability insurance.

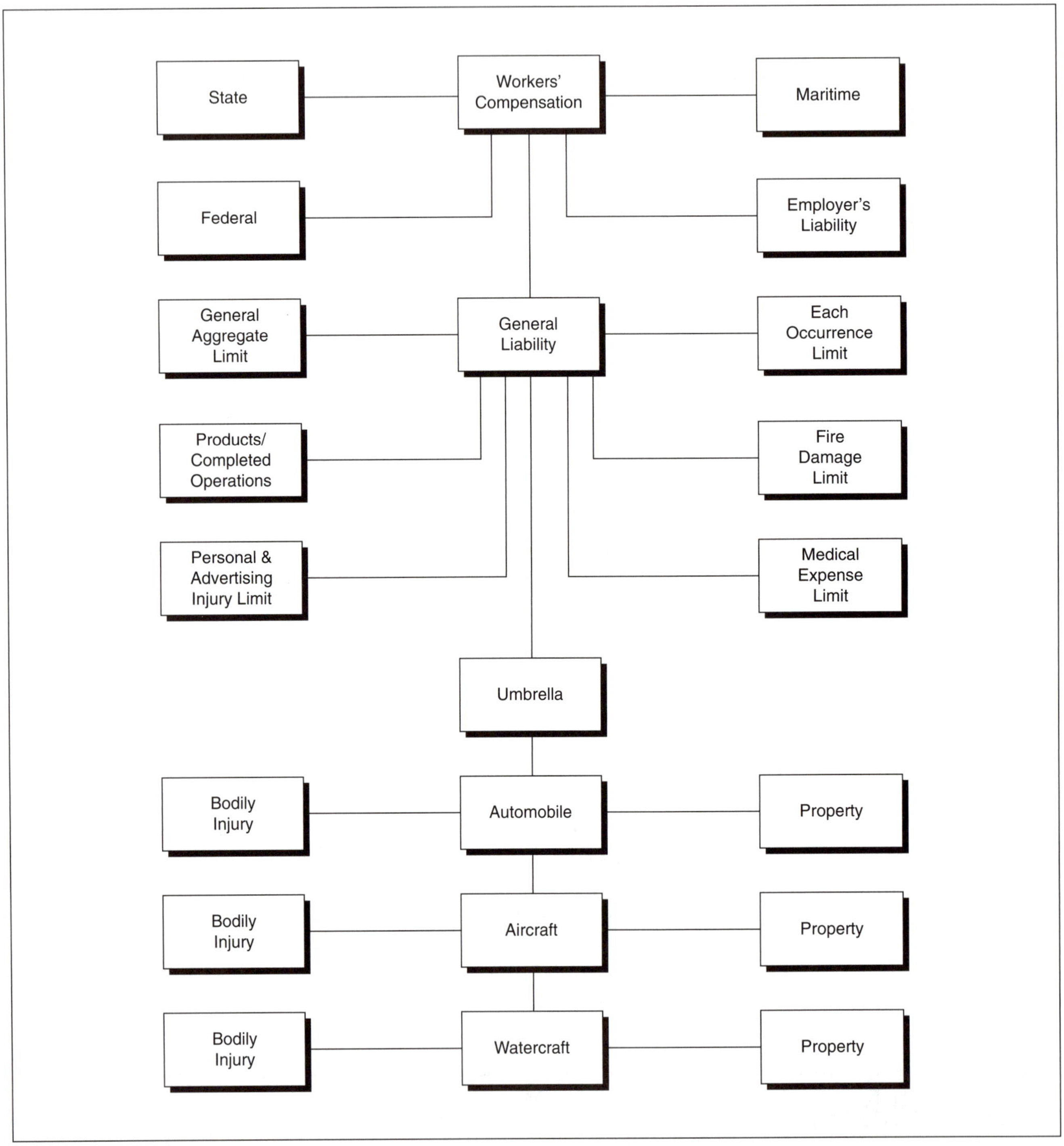

Figure 17.1
Contractor's Liability Insurance

17.3.1 Workers' Compensation Insurance

Although popularly classified as liability insurance, workers' compensation insurance is more accurately described as protection given to an employee by an employer in accordance with a statutory no-fault/limited liability agreement. Under workers' compensation laws, some form of which exists in all states, employers must compensate their employees for employment-related injuries, regardless of fault. The employee, in return, cannot bring legal action against the employer for compensation in excess of the statutory amount. Compensation includes reimbursements for medical costs and for lost wages, as well as a specific amount for certain permanent injuries. Without workers' compensation insurance, the employer does not have the protection of the state law and may be sued by the employee for unlimited damages.

When the contractor has the workers' compensation insurance required by law, the owner is also protected against suits arising from injuries to the contractor's employees. Accordingly, it is essential that the contractor and all subcontractors be required to obtain and maintain workers' compensation insurance. This requirement should be stated in the supplementary conditions. Exempt contractors should be required to carry voluntary workers' compensation insurance, which will have the same effect as that required by the law.

Some states have enacted compensation laws that are not comprehensive for every injury or every category of recoverable damages. The resulting immunity gap is covered by a policy called employer's *liability insurance.* This term is used by *AIA Document A201, General Conditions of the Construction Contract*, and *EJCDC C-700, Standard General Conditions of the Construction Contract.* Both of these general conditions require the contractor to provide this insurance in an amount deemed prudent by the owner and the owner's insurance adviser. This coverage can also extend to third parties, such as an employee's family.

Owners should also carry workers' compensation insurance, with coverage for the statutory limits, and employer's liability insurance, for protection against suits arising from injuries to their own employees. The policy should be endorsed for coverage in all states where the owner's employees may operate. The contractor and all subcontractors must comply with the laws of the states in which work for the project will be performed. This includes subassemblies fabricated in a state different from the one where the project is located. If a contractor performs work in more than one state, the contractor's insurance adviser should recommend *other states* insurance for coverage wherever the employees work. Other states insurance extends the coverage to all states listed in the endorsement schedule.

If the project will involve work related to a navigable waterway, the provisions of the United States Longshoremen and Harbor Workers Act apply. Other federal laws that may apply include the Federal Employees Liability Act and the Defense Base Act. Workers' compensation policies should include such special coverage by endorsement whenever it is needed.

17.3.2 General Liability

Under a general liability policy, often called public liability insurance or commercial general liability insurance, the insurance company agrees to pay all sums for which the insured becomes legally obligated to pay as damages. The insurance company also agrees to provide legal defense in any related suits brought against the contractor.

The terms *comprehensive general liability* and *commercial general liability* are both used to describe broad-based liability insurance. Most general liability policies were and are written on an occurrence-basis coverage. The occurrence-basis type of coverage means that the right to protection under the policy is fixed when the occurrence causing the damage is known. The policy must have been in force at the time of the occurrence, but need not have been in effect at the time the claim was made. The *claims-made* policy requires that for coverage, insurance must be continuously in force from the time of the occurrence to the time of the claim. Because the contractor normally provides the liability coverage, claims-made coverage is seldom used for construction projects because of problems with keeping the policy in force long after the project is completed. With the occurrence form, coverage that was in place during the construction period remains effective after construction has been completed for any claims resulting from the construction process.

The occurrence-basis form of insurance provides better protection for the owner. Uncertainty of coverage is a risk in a claims-made policy, as claims may be made long after the event and long after the contractor has left the project. An owner has little opportunity to monitor or enforce the contractor's maintenance of insurance coverage after the project is completed. Yet it is that future coverage on which the owner relies for protection when insurance is written on a claims-made basis. The occurrence-basis type of insurance is far easier to monitor and enforce because an occurrence usually arises at a time when the contractor is under contract with the owner.

Although an occurrence usually arises during the construction stage, an occurrence could be discovered after the work is complete and give rise to a claim. For this reason, the owner should consider requiring the contractor to provide coverage for a reasonable period following completion of the contract and require certificates of insurance as proof of that coverage. The owner should consult with a qualified insurance adviser to determine the length of that period.

AIA Document A503, Guide for Supplementary Conditions, accommodates either of the two types of general liability insurance. *AIA Document A503* uses both terms, *comprehensive* and *commercial,* for general liability insurance. It also addresses requirements if the coverage is to be provided on a claims-made basis. Either insurance industry global standard *ACORD (Association for Cooperative Operations Research and Development) Form 25, Certificate of Liability Insurance,* or *AIA Document G715, Supplemental Attachment for ACORD Certificate of Insurance 25-S,* may be used.

AIA Document G715 covers some additional subjects the ACORD form does not, such as general aggregate—this project only, certificate at final payment, and obligation to notify. *AIA Document A201* requires the contractor to file certificates of insurance, acceptable to the owner, with the owner prior to start of work. *EJCDC C-700* requires the contractor to deliver to the owner certificates of insurance and other requested evidence that the required insurance coverage has been purchased.

When required by particular construction risks, certain important additions to the general liability coverage are advisable. In its basic form, the comprehensive general liability policy has an exclusion usually called the XCU exclusion, which is an abbreviation for explosion, collapse, and underground damage. Removal of this exclusion is required to provide coverage for claims arising from blasting, collapse of structures due to excavation or removal of shoring or support, and underground damage caused by mechanical excavation. It is also in the owner's best interest to require the contractor to provide third party personal injury coverage.

The commercial general liability policy is one that combines several coverage aggregates into a single general aggregate, which is the maximum amount that will be paid under the policy. The general aggregate may be modified to apply to each individual project, and the requirement for this modification should be identified in the supplementary conditions.

Six separate limits of liability are covered in commercial general liability policies. The following is an explanation for each of the six limits as stated in the *Guide to Construction Insurance* published by the International Risk Management Institute:

1. *General Aggregate Limit.* Places an aggregate limitation on what the policy will pay for all medical expenses, personal/advertising injury, bodily injury, and property damage claims within the policy period.
2. *Products/Completed Operations Aggregate.* Places an aggregate limitation on the amount of insurance available to cover any products/completed operations claims.
3. *Personal and Advertising Injury Limit.* Constitutes the maximum insurance available to pay a claim for personal injury or advertising injury. The insurance available is further limited by the general aggregate limit.
4. *Each Occurrence Limit.* Places a limitation on the amount of insurance available to pay bodily injury, property damage, and medical expense claims arising out of any one occurrence, subject to the general policy aggregate.
5. *Fire Damage Limit.* Limits the amount of insurance available to pay for property damage (to premises rented to the insured) caused by any fire, subject to each occurrence limit as well as the general aggregate limit.
6. *Medical Expense Limit.* Represents the amount of insurance available to pay for all medical expenses arising from bodily injury sustained by any one person.

Special consideration must be given to "general aggregate limits" in the commercial general liability policy. This limit establishes the maximum payment under the policy for all claims paid under the policy. Thus, if one or more claims have been paid during the policy years, the actual aggregate limits may be less than those shown on the insurance policy or the certificate of insurance provided by the insurance agent.

17.3.3 Automobile Insurance

The beginning of a construction project is a good time for the owner to review its automobile coverage. Ownership and operation of motor vehicles is one of the more risky business operations, and the addition of a construction site to the places visited by the owner's drivers suggests the need for coverage review.

From a contractor's standpoint, the liability for operation of owned or non-owned automobiles, including rented or hired vehicles, is generally secured through a business automobile coverage form. Both AIA and EJCDC guides require the occurrence type of coverage that includes coverage for bodily injury or property damage caused by an occurrence-basis and arising out of the use of covered vehicles.

Owned automobiles are those owned by the contractor. However, non-owned automobiles can also pose a risk of liability. For example, a contractor's employee may use a personal automobile to run an errand for the contractor. Therefore, non-owned, rented, or hired vehicle coverage should be required as part of the automobile insurance coverage.

The coverage agreements and definitions in automobile liability policies uniformly exclude aircraft and powerboats exceeding a specified length and horsepower limit. Where use of aircraft or water vessels is contemplated, such as in the helicopter erection of skylights or lifting of heating, ventilating, and air-conditioning equipment, the owner should

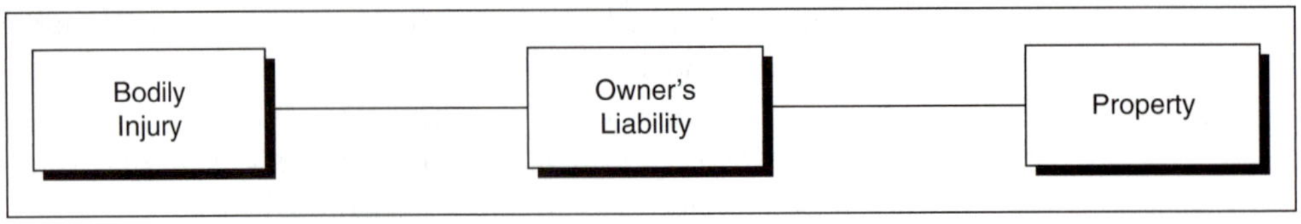

Figure 17.2
Owner's Liability

require appropriate liability coverage for both the owned and non-owned vehicles. Coverage should also be anticipated for all maritime projects and distant sites that may require air transport of material, equipment, or personnel.

17.3.4 Owner's Protective Liability

The general liability insurance policy regularly carried by the owner for normal operations may not include construction risks. Therefore, the owner's insurance coverage should be reevaluated because of exposure to construction activities. Additional coverage, called *owner's protective insurance,* may be appropriate to cover the same hazards as those covered by the contractor's general liability insurance. This is not duplicate coverage, as a legal liability exists for the owner's role in the construction process. In fact, the general conditions require the owner to obtain such coverage. It should include coverage for liability from claims arising from the contractor's operations and the owner's duties with respect to the construction project. The contractor can be required to obtain the owner's protective insurance. The owner's interests are better served, and obtaining coverage from the contractor's carrier rather than the owner's carrier minimizes conflicts between insurance policies. Figure 17.2 shows the relationships concerning owner's protective liability insurance.

17.4 Property Insurance

Property insurance compensates the insureds for damages to covered property caused by insured perils. Builder's risk insurance is the cornerstone of the project's property insurance. The relationships of the various covered risks are shown in Figure 17.3.

17.4.1 Builder's Risk Insurance

Builder's risk insurance covers losses during the construction stage arising from the insured risks of fire, windstorm, collapse, and theft. The policy is designed to cover all property that has been or will be incorporated into the project. Builder's risk insurance is available in a *named peril* or an *all-risk* type of policy. Both the AIA and EJCDC general conditions require the all-risk type of policy.

Both AIA and EJCDC general conditions state the amount of the property insurance required:

- *AIA Document A201*: "... property insurance in the amount of the initial Contract Sum as well as subsequent modifications thereto for the entire Work at the site on a replacement cost basis without voluntary deductibles."
- *EJCDC C-700*: "... in the amount of the full replacement cost thereof (subject to such deductible amounts ...)."

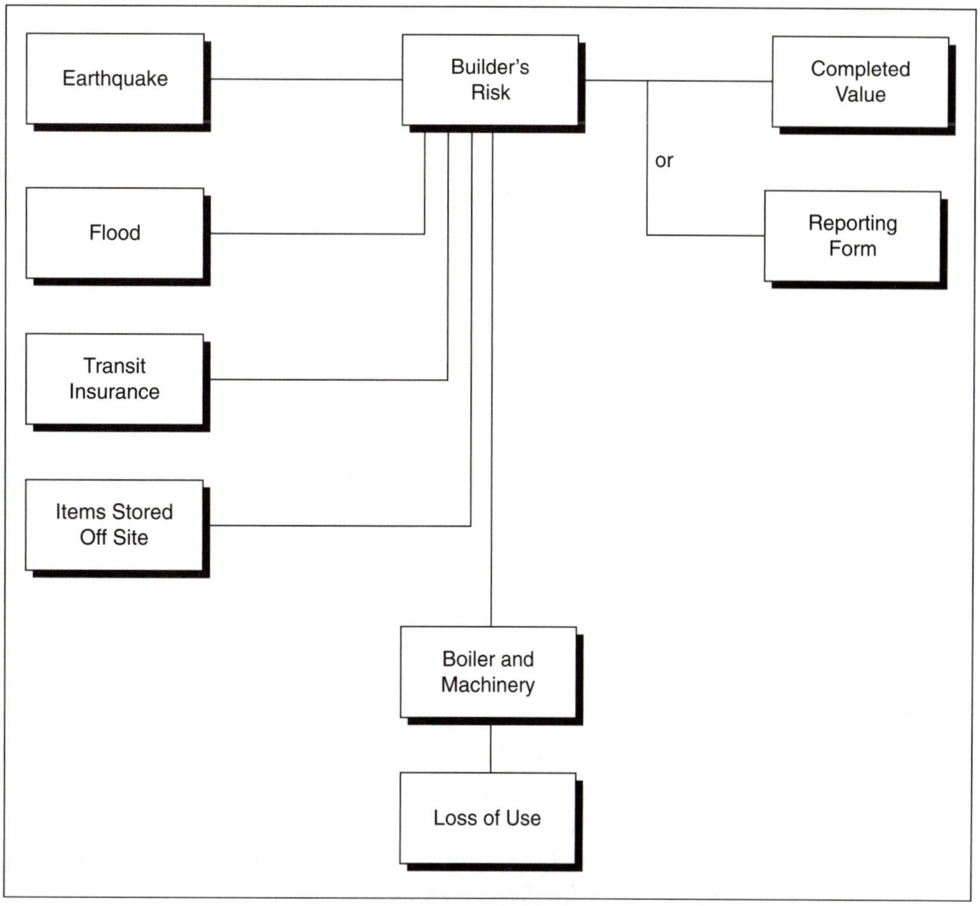

Figure 17.3
Property Insurance

Most owners prefer the all-risk form of insurance coverage, which includes coverage for all perils except those specifically excluded in the policy. The named peril type of coverage specifies the perils that are insured. All-risk is much more inclusive than the named peril type of policy. Depending on the project location, particular attention should be paid to earthquake and flood exclusions contained in some all-risk policies. Other typical exclusions may include loss of use or occupancy, penalties for noncompliance or noncompletion of the contract, normal wear and tear, latent defects, and loss caused by faulty workmanship or design.

The owner often obtains builder's risk insurance, because the owner has an insurable interest in the project. As materials are installed or stored on-site and title passes to the owner, the insured interest increases to the contract sum, to the insured limit, or until the owner occupies the project. Builder's risk insurance is available in two forms: *completed value* or *reporting*.

The completed value form is written for the anticipated completed value of the project. Most projects are suited to the use of the completed value form, especially if the owner is inexperienced in dealing with construction projects or if only one project is in progress at a time.

The reporting form is written on the actual value of the project at the time the policy is written. Therefore, the reporting form coverage must be revised on a regular basis (usually to coincide with each application for payment), with the policy premium and insured value increasing at each update. The advantage of this type of insurance is that the payment of premiums is based on insured values at a given time. This is most advantageous if the values increase sharply during the final phases of construction. The requirement for filing regular reports of values may be cumbersome to some owners; however, failure to satisfy reporting requirements can leave the owner underinsured. The reporting form works well for large projects with lengthy construction periods. An owner dealing with many projects at one time may also find the reporting form to be advantageous.

The owner should arrange for coverage of owner-furnished materials or equipment not included in the construction contract price. The owner should also determine whose insurance will cover the value of material or equipment in transit or stored by the contractor off-site during the course of the project.

- *AIA Document A201*: "… this [the owner's] property insurance shall cover portions of the Work stored off the site after written approval of the Owner, at the value established in the approval, and also portions of the Work in transit."
- *EJCDC C-700*: "Unless otherwise provided in the Supplementary Conditions, Owner shall purchase and maintain property insurance upon the Work. … This insurance shall cover materials and equipment stored at the site or at another location that was agreed to in writing by Owner … provided that such materials and equipment have been included in an Application for Payment recommended by Engineer …"

When AIA or EJCDC general conditions are used, the supplementary conditions must clearly state whether the contractor is to provide this coverage. Property of the contractor stored on the owner's premises but off the construction site should be addressed in either the owner's property insurance or the contractor's property insurance. It is sufficient for the supplementary conditions to require the party holding title to secure insurance for loss of the property. In any event, property insurance is necessary to cover loss of material whether stored on- or off-site.

In the event the owner intends to furnish material or equipment to be incorporated into the project, care must be taken to secure coverage from the time the owner assumes title. The owner's property insurance may require adjustment for the storage of these materials, while the owner has the risk of loss.

Members of the design team do not have an interest in the property, so they need not be named insured. Both the contractor and the owner have an insurable interest in the project, so either party can obtain the builder's risk policy. Normally, however, the owner will obtain this coverage. Both AIA and EJCDC general conditions require the owner to provide this insurance. Owners with limited construction experience may find administration of an insurance program burdensome, and they may be unable to obtain a ready source of coverage at favorable rates. Therefore, it may sometimes be advantageous for the owner to require that the contractor obtain the builder's risk insurance. However, owners dealing with many construction projects with the necessary administrative support in place, may prefer dealing with brokers and insurance carriers of their choice as a matter of business policy and cost control, and may obtain better rates than those available to a contractor.

The type of project may also influence the choice of who should provide builder's risk insurance. For example, an addition to an existing structure might be more economically covered by an extension of the owner's present coverage. In this type of project, the contract amount may be only a fraction of the value of the property that could be lost in the event of a major fire. In this instance, any project-related insurance must be coordinated with the owner's existing property insurance to ensure that the full value of the completed facility is protected. In addition, there may be value in the contractor obtaining special insurance coverage to protect the contractor in the event the contractor is found liable for a fire that destroyed nonproject areas in an existing building.

A completely new structure, however, may be covered more economically by the contractor's builder's risk insurance. If the construction site involves special hazards, such as chemical processing, the owner's carrier may be capable of offering better terms because of familiarity with the risks involved. If the contractor provides builder's risk insurance, the potential exists for conflicts between the owner's insurance carrier and the contractor's insurance carrier.

Owner-provided builder's risk insurance may afford cost-savings opportunities for large and complex projects. Multiple-prime contracts in which each prime contractor is required to obtain insurance may duplicate policy costs that can otherwise be saved by combining coverage into a single owner-purchased policy. For example, experience on large road construction projects involving multiple-prime contracts has demonstrated overall insurance premium savings when the owner purchases the insurance. In the case of a project with multiple-prime contracts where the owner does not provide the builder's risk insurance, the supplementary conditions should identify which of the prime contractors is to provide the insurance, and the selected prime contractor must include in the coverage the value of the other prime contracts. The contractor providing the insurance should furnish a copy (or proof of coverage) of the policy to the other prime contractors. In some cases dealing with multiple-prime contracts, the all-risk type of coverage may not be available. Regardless of who obtains the builder's risk policy, a single policy should be issued to cover the insurable interests of the owner, contractor, and all subcontractors, to avoid duplication of premiums.

Insurance coverage can be priced by the contractor and by the owner. Shopping for insurance may be feasible if the project is for a private owner or if the contract price is negotiated. The savings, if any, may compensate for the administrative demands. In the case of competitively bid public work, this approach is not feasible. The economies, if any, may compensate for the administrative demands. The project's loss performance will likely affect the future rates of the party that procures the policy. The owner, in consultation with an insurance adviser, should determine preferences.

The inclusion of many parties under a single policy requires a plan for the administration of loss sums due in part to several of the insured's. The current practice is to require the owner to adjust all losses, receive all loss payments, and distribute settlements to either the general contractor for division among the subcontractors or to each prime contractor in a multiple-prime contract project, as their interests may appear.

Subcontractors involved in the project can purchase an *installation floater*, which normally covers property only during the period of installation. The purpose of an installation floater is to insure the materials and work put in place by the subcontractor before work is completed and accepted by the owner. Installation floaters are generally required when expensive equipment or materials are involved, such as generators, chillers, or similar items. Installation floaters are usually not required if the builder's risk policy names the subcontractor and has broad-based coverage that includes equipment to be installed in the project.

Partial occupancy of a project before completion of the construction changes the insurer's risk. Therefore, partial occupancy should not begin without first obtaining the insurance carrier's permission. Regardless of who provides the property insurance, the AIA general conditions require the insurance carrier providing the property insurance to be contacted prior to occupancy and written notice to be received from the carrier stating that necessary changes in coverage have been made. The EJCDC position is that proper endorsement should be obtained so protection is ensured. Following are some common endorsements that may be added to the property insurance coverage unless all risk coverage is provided in the basic property damage policy.

- *Extended Coverage Endorsement* includes additional perils for windstorm, hail, explosion (excluding steam boiler explosion), riot, riot attending a strike, civil commotion, aircraft, vehicles, and smoke (required by both AIA and EJCDC).
- *Vandalism and Malicious Mischief* (required by both AIA and EJCDC).

- *Demolition* ordered by public authorities, also known as building and ordinance coverage (required by both AIA and EJCDC).
- *Earthquake and Flood Insurance* (both are required by AIA; only earthquake is required by EJCDC). Any deductible amounts that apply to the property insurance should be made known to the contractor. Both AIA and EJCDC recommend that deductible amounts be identified in the supplementary conditions. For cost-plus and GMP contracts, the means of charging the losses within the deductible amount should be addressed in the supplementary conditions.

17.4.2 Boiler and Machinery Insurance

Insurance companies have traditionally separated certain risks from others. One example of such separation is boiler and machinery insurance. Typically, it is a policy separate from the builder's risk policy. The boiler and machinery policy insures against loss resulting from accidents to boilers and pressure vessels identified in the policy, including damage to the property of others. When these types of equipment are part of the project, this insurance should be required. Testing of the equipment is not normally a part of the insurance, but it could be added by an endorsement, and the owner should consider the need for such an endorsement. In addition, the insured equipment is not covered during hoisting operations.

Both the AIA and EJCDC general conditions state that the owner is to provide boiler and machinery insurance, when such equipment is included in the project. If the owner wishes the contractor to provide the insurance, the supplementary conditions must identify the equipment to be covered and the limits of insurance.

17.4.3 Contractor's General Property and Equipment Insurance

The builder's risk policy covers the value of the construction, but it does not cover the value of the tools and equipment used to build the project. Loss of tools and equipment can affect the progress and, perhaps, the quality of the work. Therefore, the owner should require the contractor to obtain insurance covering construction equipment and tools used on the project or show the financial ability to replace them.

17.4.4 Business Interruption Insurance

An accident during the course of construction, such as a fire, may prevent the owner from obtaining use of the project by the planned completion date. Financial loss and additional expense may result. To guard against this possibility, an owner may obtain loss-of-use insurance that protects against loss caused by the inability to occupy and use the project because of damage resulting from a covered peril. This coverage is available for a variety of risks, including fire, and can include compensation for additional business expense and for the rental of substitute facilities. The owner may elect to have this coverage throughout the period of construction, and because of the untested performance of a new facility, extend the coverage for a period following construction. *AIA Document A201* states that the owner, at the owner's option, may purchase and maintain such insurance. This coverage can be written as an extension of the builder's risk policy and insures for loss of use arising from the same perils as those insured against under the builder's risk coverage.

17.4.5 Umbrella Excess Liability Insurance

The umbrella excess liability policy provides a layer of insurance above the limits carried in the contractor's other liability coverage. Two purposes are served by specifying a relatively high limit umbrella excess liability policy. First, this is the most economical way to secure limits in excess of those normally offered by general liability and automobile policies. Second, the umbrella can broaden the coverage provided by those policies.

AIA Document A503 addresses this policy by requiring a dollar amount of primary insurance and a dollar amount of retention for each occurrence of self-insured hazards. In *EJCDC C-800,* it is addressed by requiring a dollar amount for the general aggregate and for each occurrence.

17.4.6 Subrogation

The objective of comprehensive insurance planning is to create a network of policies to protect the collective and separate interests of parties engaged in the construction project. Subrogation in favor of insurance companies is a right that can greatly disrupt the relationships of the parties involved.

Subrogation allows the insurer to assume the insured party's rights against others in exchange for payment of the loss or damage. For example, suppose an owner suffers losses resulting from fire when a new electrical transformer malfunctioned. The owner is insured against fire by insurance company "A" and files a claim for the fire damage. "A" investigates and pays the full loss. However, in the investigation, "A" determines that the transformer was not installed as required by the manufacturer's instructions or that it was improperly designed by the electrical engineer. With the loss having been paid, the subrogation clause takes effect, and insurance company "A" obtains the rights to all the claims the insured (owner) could bring against the other parties for the damages. The insurance company then becomes a subrogee that can now file suit against either the contractor or the engineer as appropriate to satisfy the claim.

Through subrogation, one party's settlement can become another party's lawsuit. Because of the threat of litigation, this situation may adversely affect the owner's relationship with the contractor and the A/E for the balance of the project. To prevent this situation, current contract practice suggests that all parties involved in the project waive all claims against one another to the extent covered by insurance. Because of this waiver, the right of subrogation does not arise. All parties involved in the construction contract should check whether such a waiver violates any term or condition of their individual insurance policies.

17.5 Comparison of Insurance Coverage

A comparison of the text in the AIA and EJCDC General Conditions regarding insurance requirements is shown in Figure 17.4.

AIA Document A503 provides commentary along with suggested modifications for the insurance portion of the AIA General Conditions. *EJCDC C-800* provides commentary and suggestions for modifying the insurance portion of the EJCDC general conditions. These documents establish concepts and requirements that set a standard for the insurance coverage that should be provided for construction projects. However,

AIA Document A201, *General Conditions of the Contract for Construction* (2007 Edition)	**EJCDC C-700, *Standard General Conditions of the Construction Contract* (2007 Edition)**
ARTICLE 11 INSURANCE AND BONDS	**ARTICLE 5—BONDS AND INSURANCE**
11.1 CONTRACTOR'S LIABILITY INSURANCE **11.1.1** The Contractor shall purchase from and maintain...such insurance as will protect the Contractor...	5.04. *Contractor's Liability Insurance* A. Contractor shall purchase and maintain such liability and other insurance as is appropriate for the Work being performed...
11.1.2 ...written on an occurrence or claims-made basis...maintained without interruption from date of commencement of the work until date of final payment...	... B.3. include contractual liability insurance covering Contractor's indemnity obligations under Paragraphs 6.11 and 6.20;
...	B.4. contain a provision or endorsement that the coverage afforded will not be canceled, materially changed or renewal refused until at least 30 days prior written notice has been given to Owner and Contractor and to each other additional insured identified in the Supplementary Conditions to whom a certificate of insurance has been issued (and the certificates of insurance furnished by the Contractor pursuant to Paragraph 5.03 will so provide);
11.2 OWNER'S LIABILITY INSURANCE **11.2.1** The Owner shall be responsible for purchasing and maintaining... ...	... 5.05. *Owner's Liability Insurance* A. ...all Owner's option may purchase and maintain at Owner's expense Owner's own liability insurance...
11.4 PROPERTY INSURANCE **11.4.1** ...the Owner shall purchase and maintain...property insurance written on a builder's risk "all risk" or equivalent policy form in the amount of the initial Contract Sum, plus value of subsequent Contract modifications... for the entire Project...	5.06. *Property Insurance* A. ...Owner shall purchase and maintain property insurance upon the Work at the site in the amount of the full replacement cost...
11.4.1.1 Property insurance shall be on an "all-risk" or equivalent policy form and shall include, without limitations, insurance against the perils of fire (with extended coverage) and physical loss or damage including, without duplication of coverage, theft, vandalism, malicious mischief, collapse, earthquake, flood, windstorm falsework, testing and startup, temporary buildings and debris removal including demolition occasioned by enforcement of any applicable legal requirements...	A.2. be written on a Builder's Risk "all-risk" or open peril or special causes of loss policy form...
11.4.1.2 If the Owner does not intend to purchase...the Owner shall so inform the Contractor...	
11.4.1.3 If the property insurance requires deductibles the Owner shall pay cost not covered...	
11.4.1.4 This property insurance shall cover portions of the Work stored off the site and also portions of the Work in transit. ...	A.4. cover materials and equipment stored at the Site or at another location that was agreed to in writing by Owner prior to being incorporated in the Work, provided that such materials and equipment have been included in an Application for Payment recommended by Engineer.
11.4.2 Boiler and Machinery Insurance. The Owner shall purchase and maintain...	B. Owner shall purchase and maintain such boiler and machinery insurance or additional property insurance as may be required.... ...
11.4.3 Loss of Use Insurance. The Owner, at the Owner's option, may purchase and maintain...	D. Owner shall not be responsible for purchasing and maintaining any property insurance specified in this Paragraph 5.06 to protect the interests of Contractor... ...
11.4.7 Waivers of Subrogation. The Owner and Contractor waive... ...	5.07 *Waiver of Rights* A. ...Owner and Contractor waive all right against each other..
9.9 PARTIAL OCCUPANCY OR USE **9.9.1 The Owner may occupy or use**... provided such occupancy or use is consented to by the insurer...	5.10. *Partial Utilization, Acknowledgment of Property Insurer* A. ...no such use or occupancy shall commence before the insurers providing the property insurance...have acknowledged notice thereof and in writing effected any changes in coverage necessitated thereby. The insurers providing the property insurance shall consent by endorsement on the policy or policies, but the property insurance shall not be canceled or permitted to lapse on account of any such partial use or occupancy.

Figure 17.4
Comparison of AIA and EJCDC Text

each project should be reviewed by the owner's legal counsel and insurance adviser to determine specific needs.

17.5.1 The Additional Insured

A liability policy protects persons named on the policy agreement. These persons are called *named insureds.* When so named, a person is entitled to a legal defense for covered claims and to policy payment of damages on the insureds' behalf for the legal liability arising from a covered occurrence.

When injury or damage results from the contractor's construction operations, other parties, together with the contractor, can be exposed to lawsuits. The phenomenon called the *shotgun lawsuit* can draw numerous parties, including the owner and the A/E, into litigation. The intention may be to sort out the responsible parties and thereby serve the interests of justice, but the sorting process is expensive for the named parties involved only in the project but not in the damage.

The conditions of the contract should require the owner and the A/E to be named as additional insureds under the contractor's liability coverage. It must be understood that the insurance company does not expand the coverage of the policy by that extension. An additional insured is merely afforded the benefits of the policy for the operations of the contractor (the named insured). The acts, errors, or omissions of the owner and the A/E are not covered under the contractor's policy. Errors and omissions coverage must be provided by the individual firm's professional liability policy.

This additional insured endorsement to the contractor's policy is not a substitute for the owner's protective liability insurance. The coverage by an owner's policy will be greater than that obtained by being named an additional insured in the contractor's policy.

17.6 Obtaining and Incorporating Information

The general conditions published by AIA and EJCDC do not contain dollar amounts of insurance values to be used on the project. The proper place to include insurance values is in the supplementary conditions because the requirements vary from one project to the next. The actual coverage amounts must be obtained from the owner and inserted in the appropriate articles. This information should be requested early enough to allow the owner time to consult its insurance adviser and reply to the A/E in sufficient time for the information to be incorporated into the supplementary conditions, in the instance when they are being prepared by the A/E. When performing work for a repeat client or with an owner constructing multiple projects, the A/E should request insurance information for each individual project.

The insurance instruction request forms published by AIA or EJCDC are recommended for documenting the transfer of insurance information from the owner to the A/E. These forms also provide the A/E with a standardized means of ensuring that all necessary information has been requested and obtained. The AIA form is *AIA Document G612, Owner's Instructions to the Architect,* regarding the construction contract, insurance and bonds, and bidding procedures. The EJCDC form is *EJCDC C-052, Owner's Instructions Concerning Bonds and Insurance for Construction.* In addition, the EJCDC also publishes a prototype letter for requesting the information. This document, *EJCDC C-051*, is titled *Engineer's Request for Instructions on Bonds and Insurance.*

The supplementary conditions can be treated the same as a master guide specification section with spaces left blank for insurance information and other modifications as recommended by either AIA or EJCDC and used in the A/E's practice. The A/E should use the language recommended in either the AIA or EJCDC guides when preparing supplementary conditions. The portion of the supplementary conditions dealing with insurance information should be identical to the request form used and should include blank spaces where information will be transferred from the request form to the supplementary conditions. Upon receipt of the completed *AIA Document G612* or *EJCDC C-052* from the owner, the A/E should insert the insurance information into the appropriate place in the supplementary conditions. If the form has missing information or appears to be incorrectly completed, the A/E should send it back to the owner, requesting clarification or completion. The A/E should not try to guess the intent of the owner. The A/E is not the owner's insurance adviser, nor does the A/E's liability insurance provide coverage for specifying insurance information.

The addition of insurance values is a normal modification to the general conditions. Several items regarding insurance should be addressed:

- *Aircraft Liability.* Add wording to the supplementary conditions when directed by the owner.
- *Watercraft Liability.* Add wording to the supplementary conditions when directed by the owner.
- *Owner's Liability Insurance.* If the contractor is required by the owner to provide this coverage, modify the general conditions with wording in the supplementary conditions.
- *Property Insurance.* If the contractor is required by the owner to provide the property insurance, major modifications to the general conditions must be made in the supplementary conditions.
- *Builder's Risk Insurance.* If the project consists of multiple-prime contracts, the owner should assign the responsibility for builder's risk coverage to only one of the prime contractors.
- *Boiler and Machinery Insurance.* If necessary, modify general conditions in the supplementary conditions.
- *Work Stored Off the Site.* If project requirements are different from those in the general conditions, modify general conditions in the supplementary conditions.

17.7 Glossary of Insurance Terms

The following definitions related to construction insurance are taken from the *Glossary of Insurance and Risk Management Terms* published by the International Risk Management Institute (IRMI).

All-Risk Insurance Protection from loss arising out of any fortuitous cause other than those perils or causes specifically excluded by name. This is in contrast to other policies that name the peril or perils insured against. Refer to *Named perils.*

All-Risk—Difference in Conditions (Builder's Risk) In many cases, the owner or general contractor, as in the case of a subcontractor, will carry builder's risk insurance on a particular project. Many times these builder's risk programs may have more restrictive coverage compared with what the general contractor or subcontractors may ordinarily purchase. Also, these programs usually are built around substantial deductibles. Therefore, it is desirable to have a builder's risk installation policy that is a difference-in-conditions policy. For any loss that is excluded or within the deductible of the owner or general contractors program, coverage may be provided by the difference-in-conditions program. In most cases, the difference-in-conditions rate is roughly 40 to 50 percent of the annual builder's risk rate.

Boiler and Machinery Insurance Insurance against loss arising from the operation of boilers and machinery. May cover loss suffered by the boilers and machinery or include damage done to other property and business interruption losses.

Builder's Risk Indemnifies for loss of or damage to a building under construction from specified perils. Insurance is normally written for a specified amount on the building and applies only in the course of construction. Coverage usually includes fire and extended coverage and vandalism and malicious mischief coverage. Builder's risk coverage can be extended to an all-risk form as well. The builder's risk policy also may include coverage for items in transit to the construction site (up to a certain percentage of value) as well as items stored at the site.

Claim Used in reference to insurance, a claim may be a demand by an individual or corporation to recover, under a policy of insurance, for loss that may come within that policy.

Claims-Made Policy Policies written under a claims-made basis will cover claims made (reported or filed) during the year the policy is in force for any incidents that may occur that year *or* during any previous period the policyholder was insured under the claims-made contract. This form of coverage is in contrast to the occurrence policy that covers today's incident regardless of when a claim is filed, even one or more years later.

Completed Operations A form of liability insurance that provides coverage for bodily injury and property damage arising out of the operations that have been completed or abandoned, provided the accident occurs away from the premises owned or rented by the insured. Operations shall be deemed completed at the earliest of the following times:

- When all operations to be performed by or on behalf of the insured under the contract have been completed.
- When all operations to be performed by or on behalf of the insured at the site of the operations have been completed.
- When the portion of the work from which the injury or damage arises has been put to its intended use by other than the contractor or subcontractor.

Comprehensive General Liability (CGL) Policy A broad form of liability insurance usually covering business organizations to protect them against liability claims for bodily injury and property damage arising from ongoing operations, products and completed operations, and independent contractors, but excluding coverage for liability arising from the use of automobiles.

Comprehensive Policy This term applies to a variety of policies that provide broad protection.

Contractual Liability Liability assumed under any contract or agreement over and above that liability that may be imposed by law.

Coverage In insurance practice, the word *coverage* is used synonymously with *insurance* or *protection.*

Employer's Liability Insurance This coverage is provided by Part 2 of the basic workers' compensation policy and pays on behalf of the insured (employer) all sums that the insured shall become legally obligated to pay as damages because of bodily injury by accident or disease sustained by any employee of the insured arising out of and in the course of employment by the insured.

Endorsement A form bearing the language necessary to record a change in an insurance policy.

Exclusion A provision of an insurance policy or bond referring to hazards, circumstances, or property not covered by the policy.

General Aggregate Limit The maximum limit of liability payable during any given annual policy period by an insurer under the 1986 CGL policy on behalf of an insured for all losses other than those arising from the products and completed operations hazards.

Hold-Harmless Agreement A contract under which legal liability of one party for damages is assumed by the other party to the contract. The basic types of such agreements are as follows:

- Limited form, where Party A reaffirms responsibility for Party A's own negligent acts; Party B is thus protected when Party B is held vicariously responsible.
- Intermediate form, where Party A reaffirms responsibility for Party A's own acts and agrees to share responsibility for joint and concurrent negligence of both parties.
- Broad form, where Party A reaffirms responsibility for all liability, including that arising out of the sole negligence of Party B.

Indemnify To restore in whole or in part the victim of a loss by payment, repair, or replacement.

Indemnity Restoration to the victim of a loss up to the amount of the loss.

Indemnity Clause See *Hold-harmless agreement.*

Installation Floater Usually written to cover machinery and equipment of all kinds during transit, installation, and testing at the purchaser's premises.

Insurable Interest An interest by the insured person in the value of the subject of insurance, including any legal or financial relationship. Insurable interest usually results from property rights, contract rights, and potential legal liability.

Insurance A contractual relationship that exists when one party (the insurer), for a consideration (the premium), agrees to reimburse another party (the insured) for loss to a specified subject (the risk) caused by designated contingencies (hazards or perils). The term *assurance,* commonly used in England, is ordinarily considered identical to, and synonymous with, insurance.

Insured The person(s) protected under an insurance contract.

Liability Insurance Insurance paying or rendering service on behalf of an insured for loss arising out of legal liability to others.

Limit of Liability The maximum amount that an insurance company agrees to pay in the case of loss.

Longshoremen and Harbor Workers Act A federal act requiring certain compensation levels for injured longshoremen and harbor workers. This exposure may be insured commercially or by using self-insurance. The classifications of persons falling under the provisions of this act are broadening widely.

Loss (1) The basis of a claim for damages under the terms of a policy. (2) Loss of assets resulting from a pure risk. Broadly categorized, the types of losses of concern to risk managers include personnel loss, property loss, time element, and legal liability.

Loss-of-Use Insurance Coverage to compensate the policyholder for loss suffered because the policyholder cannot use the property destroyed or damaged by an insured peril.

Monopolistic Fund States State organizations that have the exclusive right to insure within the respective states the employer's workers' compensation liability. The monopolistic fund states are Nevada, North Dakota, Ohio, Oregon, Washington, West Virginia, and Wyoming.

Named Insured Any person, company, or corporation or any of its members specifically designated by name as insured(s) in the policy as distinguished from others who, although unnamed, are protected by the policy definition. A named insured under the policy has rights and responsibilities not attributed to additional insureds, such as premium payment, premium return, notice of cancellation, and dividend participation.

Named Perils Named peril policies specify the perils that are insured against, as distinguished from the all-risk and broad-form policies.

Occurrence An event that causes injury to people or damage to property.

Occurrence (CGL) This term means an accident, including continuous or repeated exposure to conditions, that results in bodily injury or property damage neither expected nor intended from the standpoint of the insured. Occurrence policies cover claims made that arise from injury or damage that occurs during the policy period, irrespective of when the claim is made against the insured. See *Claims-made policy.*

Other States Coverage Workers' compensation and employers' liability insurance afford coverage only for the states listed on the information page. For insureds that may have employees traveling through other states, the endorsement may be used. By adding this endorsement to the policy, coverage is extended to all state laws listed in the endorsement's schedule. However, coverage cannot be extended in this manner to monopolistic fund states.

Owner's and Contractor's Protective Liability (OCP) Coverage for bodily injury or property damage liability caused by an occurrence and arising from the operations performed for the named insured by the contractor designated, or acts or omissions of the named insured in connection with the named insured's general supervision of such an operation. Coverage can be obtained by separate policy. The CGL policy provides the insured with automatic coverage for liability arising from independent contractors.

Peril The cause of a loss insured against in the policy (e.g., fire, windstorm, explosion).

Policy The insurance policy is a contract or agreement between the insurer and the insured.

Policy Term The period for which the premium has been paid.

Property Damage As defined in the general liability policy, property damage means (1) physical injury to or destruction of tangible property that occurs during the policy period, including the loss of use thereof at any time there from, or (2) loss of use of tangible property that has not been physically injured or destroyed, provided such loss of use is caused by an occurrence during the policy period.

Property Insurance Insurance that indemnifies the person with an interest in the physical property for its loss or the loss of its income-producing ability. This is first-party insurance, which provides coverage for the insured's property damaged or destroyed by an insured peril, as contrasted with liability insurance, which covers the insured's legal liability to others. Examples of property insurance include builder's risk and fire insurance.

Replacement Cost The replacement value of the damaged property without deduction for depreciation.

Reporting Form A policy designed for use when values of the insured property fluctuate during the policy term. Usually an adequate limit of liability is set, and then the insured reports the values actually on hand on a given day of each month. At the end of the year or policy term, these reported values are averaged and the premium adjusted accordingly.

Subrogation The assignment to an insurer by terms of the policy or by law, after payment of a loss, of the rights of the insured to recover the amount of the loss from one legally liable for it.

Use and Occupancy Insurance A term used to describe business interruption coverage that is caused by boiler and machinery occurrences. Use and occupancy coverage can be added to a boiler and machinery policy by endorsement.

Waiver of Subrogation The relinquishment by an insurance carrier of the right to collect for damages paid on behalf of the policyholder.

Workers' Compensation All states have laws that provide compensation to a worker if the worker is injured while at work for an employer, whether or not the employer has been negligent. The workers' compensation laws apply to all individuals, except those specifically excluded.

XCU Exclusion (not defined by International Risk Management Institute [IRMI]). A common exclusion from liability coverage for damages to others caused by blasting and explosion, collapse of structures, and underground excavation damage to property.

Chapter 18 Summary

The successful construction of a project results when the participants know and understand their roles, relationships, and requirements. When participants agree upon and are familiar with each other's and their own rights and responsibilities for performing aspects of the work, a sense of individual purpose and team cooperation can prevail over the project.

To help in the dissemination of the project requirements, a set of documents is developed to define the procurement and contracting requirements. Of the contracting requirements, the agreement legally binds the other documents into a package of contract documents that describe the rights, responsibilities, and relationships of the parties involved, as well as the scope of the work contracted for. The contract documents include contracting forms, project forms, conditions of the contract, specifications, contract drawings, revisions, clarifications, and modifications; excluding the drawings not in an A4 (8.3×11.7 inch) format, they are bound together into a volume called the project manual.

Depending on the selected project delivery method and the contract basis of payment, the project manual and the drawings can be developed to provide for procurement and contracting. In addition to the procurement and contracting requirements, specifications establish the specific project requirements.

Division 01—General Requirements specify administrative, procedural, and performance requirements that govern the work and specification sections of the other divisions. The technical specifications can be documented using various methods. The four methods of specifying (*descriptive, performance, reference standard, and proprietary*), or any combination of the four, can be used to achieve certain construction results, except it is not recommended to combine descriptive and performance methods.

For example, the descriptive specification will provide the contractor with a detailed description of the characteristics and requirements. A performance specification will specify the desired outcome to be achieved, which affords the contractor more opportunity to use innovation, optional materials, and techniques to achieve the desired results. Other specification sections may specify basic materials and workmanship in accordance with industry-standard references. If required by the owner or a unique design function, a sole source or proprietary specification could be used. Effective specifications may include a combination of methods of specifying to establish the requirements for each portion of the work.

For efficient communication, the architect/engineer (A/E) must know how to prepare specifications that state requirements in one location in a *clear, concise, complete, and correct* manner. This requires knowledge on the part of the A/E or specifier in using proper specification language and formats to convey the project requirements to the contractor without ambiguity. In addition, to minimize discrepancies, coordination of the construction documents must be performed. To this end, rather than prepare specifications from scratch, many A/Es have developed or utilize master guide specifications that are coordinated with standard procurement and contracting requirements, and with Division 01 of the specifications. Master guide specifications can be developed from scratch or from commercially available base specification systems.

To assist the A/E in developing construction documents that are consistent and easily used, standardized formats for locating construction information have been developed. Such standard formats include *UniFormat™*, *MasterFormat®*, *SectionFormat™*, and *PageFormat™*. In addition, these standard formats are useful for helping contractors locate information that may have been otherwise difficult to find. Standardized formats for drawings have been developed as well and include the NCS, or National CAD Standard, for describing drawing content and providing a standard sheet layout.

Standardized specification and drawing formats can make the process of coordinating drawings and specifications more effective and efficient through the concept of providing a piece of information just once, and in a predetermined location, properly integrated with the other construction documents.

Construction activities can place parties at risk, so knowledge of construction bonds and insurance is a practical necessity. Construction bonds are of three basic types: *bid bond, performance bond,* and *payment bond.* The first is used during procurement, and the last two are assurances for performance and payment that might be required as a part of the contract documents. Other types of bonds such as *warranty bonds* and *maintenance bonds* may be required and are generally specified in the appropriate specification section. There are also different forms of insurance covering various aspects of construction, such as *worker injury, damage to equipment,* and *damage to the work.* The use and form of both bonds and insurance must always be determined by the owner with project specific recommendations from the owner's insurance adviser and legal counsel.

Questions and changes develop during procurement and construction of a project, which require adequate procedures for dealing with the necessary revisions, clarifications, and modifications.

Responses should be issued in written form and by using the appropriate document at the proper time in the project (i.e., addenda during the bidding/negotiation/purchasing stage and contract document modifications during the construction stage).

Special variations in documents might be appropriate to deal with specific types of projects. For instance, shortform specifications might be used for negotiated contracts where a concise, streamlined description of the work might be appropriate. Performance specifications, as described in Chapter 12, "Performance Specifying," might be utilized in a request for proposal (RFP) for design-build project delivery. Purchasing documents and specifications could be used where the purchase of goods is to be described. The federal government and other public agencies may have specific requirements and procedures regarding development and production of documents and specifications.

Correctly developed and coordinated construction documents can be utilized, along with subsequent modifications, for procurement, permitting, construction, and facility management; and can also facilitate a future reuse, renovation, or demolition of the building.

References

Derk, Walter T., *Insurance for Contractors,* 6th ed. Chicago: Fred S. James and Co., 1989.

Garner, Bryan A., ed. *Black's Law Dictionary,* 7th ed. St. Paul, MN: West Group, 1999.

International Risk Management Institute, *Insurance for Subcontractors.* Dallas, TX: International Risk Management Institute; and Alexandria, VA: American Subcontractors Association, 1984.

International Risk Management Institute, in cooperation with Associated General Contractors of America, *Guide to Construction Insurance.* Dallas, TX: International Risk Management Institute; and Alexandria, VA: American Subcontractors Association, 1985.

McIntyre, W. S., ed., *Glossary of Insurance and Risk Management Terms,* 4th ed. Dallas, TX: International Risk Management Institute, 1986.

Rothschild, Bernard B., *Construction Bonds and Insurance Guide,* 3rd ed. Washington, DC: American Institute of Architects Press, 1988.

Uniform Commercial Code, jointly drafted by the National Conference of Commissioners on Uniform State Laws, 211 East Ontario Street, Suite 1300, Chicago, IL 60611, and the American Law Institute, 4025 Chestnut Street, Philadelphia, PA 19104.

Woodward, W. J., Farmer, M. C., McIntyre, W. S., and Gibson, J. P., *Construction Risk Management,* 2 vols. Dallas, TX: International Risk Management Institute, 1984.

Index

G

H

I

J

L

M

Q

R

About the Companion Web Site

Additional CSI publications that complement *The CSI Construction Specifications Practice Guide* are available for download. The following documents can be obtained at www.wiley.com/go/csipracticeguides:

- MasterFormat® numbers and titles
- UniFormat™
- SectionFormat™/PageFormat™
- Sample CSI Forms
- GreenFormat™ questionnaire
- Practice Guide Glossary

To access the content on this site, you will need the unique Access Code printed on the card included with this book and follow these steps:

1. Navigate your web browser to: www.wiley.com/go/csipracticeguides
2. Click on the Download Access link for *The CSI Construction Specifications Practice Guide* and follow the instructions on the page.
3. Enter your access code once to set up your account and access the material.

If you have purchased an ebook version of this title or if your access code is not working, please contact Customer Care at 877-762-2974 or at http://support.wiley.com for a unique Access Code in order to take advantage of the web site.